Significance tests

Applied Statistics Series

General Editors: Geoffrey Wyborn, Kingston Polytechnic and Philip Taylor, Garnett College

Joan Macfarlane Smith, *Interviewing in market and social research*

Significance tests

Evelyn Caulcott
Department of Sociology, University of London Goldsmiths' College

Routledge & Kegan Paul   London and Boston

First published 1973
by Routledge & Kegan Paul Ltd
Broadway House, 68–74 Carter Lane,
London EC4V 5EL and
9 Park Street, Boston, Mass. 02108, U.S.A.
Printed in Great Britain by
Unwin Brothers Limited
The Gresham Press, Old Woking, Surrey, England
A member of the Staples Printing Group

ISBN 0 7100 7406 9

Preface and acknowledgments

Whilst forming part of a series of statistical textbooks, this volume is intended to stand on its own as a work dealing comprehensively with significant tests at about the first/second year course level for first degree students in sociology, economics and business studies, and also for students taking professional examinations, such as those of the ICMA and ICA. The development of theoretical ideas is therefore complete.

I have assumed therefore, that readers of this book understand any simple algebraic notation used, and that they have the necessary basic mathematical knowledge to deal with the actual calculations involved. Similarly, I have assumed that readers understand the basic descriptive measures, primarily mean and standard deviation, which are used in this book, and are able to calculate them.

In its turn, this book is intended to help those whose primary interest is in the theory of sampling and the interpretation of sample surveys. It sets out the basis of statistical inference essential to readers of *Sample Survey Methods* by G. R. Wyborn. It provides the theoretical background necessary to the critical reading of published surveys, from the viewpoint of certain statistical aspects of survey design and inference from survey results. An understanding of sampling theory in relation to inference is essential for the practioner of survey techniques.

Chapters 1 and 2 are preliminary in that they deal with points of importance to sampling theory, a grasp of which is essential before approaching sampling theory itself. In chapter 3, relationships between the population characteristics, and the characteristics of samples drawn from the population are studied. These relationships are developed in chapter 4, to inferring the population characteristics from a single sample, since significance tests cannot be discussed without dealing first with problems of estimation.

In chapter 5, this is extended to comparing the characteristics of sample and population, or of two samples, in order to decide whether one numerical measure can be said to differ significantly from the other. If this is the conclusion, then the observed difference is one that is unlikely to be due to chance, that is, to random sampling variation. For this reason, it is referred to as being 'statistically significant'. The title of this book is the same as for chapter 5, namely *Significance Tests*, because from the point of view of students dealing with sampling problems, it is this aspect of statistical inference which usually appears most urgent.

The arguments of chapter 5 are extended further in chapter 6, in examining the magnitude of the difference between two sample statistics, and the conclusions that can be drawn from this difference. This chapter exceeds the usual requirements of most first/second year courses, but it is the logical extension of chapter 5, and is therefore included in order to complete the theoretical argument. As a separate chapter it can easily be omitted by students who do not wish to pursue this aspect.

Finally, chapter 7 deals with tests for goodness of fit and association, which permit examination of the patterns of differences between a series of values. Tests are not restricted to two values only, as in chapter 5.

I have developed and expressed the argument and theory both verbally and mathematically. Thus the student whose previous experience includes little that is mathematical is not handicapped in his understanding of the underlying theory. I attach great importance to the ability to derive from verbal statements a numerical implication, and from a numerical result, a verbal interpretation. My own experience in teaching sociology students is that this dual relationship between the verbal and the numerical is something that many find difficult to comprehend and use. However, a grasp of this relationship is essential, if a proper understanding of statistical methods is to be acquired. From the student's point of view, once a clear understanding of this relationship has been achieved, a statistical methods course becomes much more satisfying, both intellectually and practically.

Finally, there are a number of acknowledgments which it is my pleasure to make.

I am grateful to Professor E. S. Pearson, as joint editor, and on behalf of the *Biometrika* Trustees, for permission to reproduce the material in Table I of the Appendix, which is taken from *Biometrika Tables for Statisticians*, Vol. 1 (3rd ed. 1966) edited by E. S. Pearson and H. O. Hartley.

I am indebted to the Literary Executor of the late Sir Ronald A. Fisher, FRS, to Dr Frank Yates, FRS and to Oliver &

Boyd, Edinburgh, for permission to reprint Tables II and III of the Appendix from their book, *Statistical Tables for Biological, Agricultural and Medical Research* (6th ed., 1963).

On a more personal basis, my thanks are due to my sometime colleague, Geoffrey Wyborn, for suggesting the writing of this book, and for his encouragement and advice, which I have greatly appreciated. I am grateful to Magda Matejewska for typing the manuscript, a task which she undertook with cheerfulness and efficiency. Last of all, my thanks go to my husband, for his encouragement, and to my children, for tolerating the upheavals attendant on authorship.

To Tom, Celia and Stephen

Contents

Glossary of symbols and formulae

Chapter 1

p indicates the probability of an event occurring.

p_1, p_2, etc. used with suitable subscripts to differentiate between events.

Chapter 2

$p(\text{H T})$ probability of obtaining results denoted by letters (indicating events) within brackets.

$[p+(1-p)]^n$ form of binomial distribution used in probability situations; p represents probability of 'success', $(1-p)$ probability of 'failure', and n, the number of 'trials'.

$\binom{n}{r}$ general form of binomial coefficient relating to r successes in n trials.

$\binom{n}{r}$ $= \dfrac{n!}{r!(n-r)!}$ and

$n!$ ('*n*-factorial') $= \text{n} \times (n-1) \times (n-2) \times \ldots \times 3 \times 2 \times 1$.

μ (pronounced mew) arithmetic mean of a set of data.

σ (pronounced sigma) standard deviation of a set of data. [Conventionally, Greek letters are used to represent *population* parameters, and Roman letters to represent *sample* statistics.]

z or z value standard units value of any variable when converted into a difference from the mean in units of standard deviation.

standard normal distribution a normal distribution with mean of 0, standard deviation of 1, and total area under the curve equal to 1.

np	mean of binomial distribution (see above).
$\sqrt{\{np(1-p)\}}$	standard deviation of binomial distribution.
p	mean proportion of binomial distribution.
$\sqrt{\{p(1-p)/n\}}$	standard deviation of proportion.
$\neq$	is not equal to.
$p \neq \frac{1}{2}$, etc.	p is not equal to $\frac{1}{2}$ (or other stated value).

Chapter 3

n	sample size.
$\sigma/\sqrt{n}$	standard deviation of theoretical sampling distribution (a normal distribution) of means of samples, size n, when drawn from a population with standard deviation, σ.
θ (pronounced theta)	Greek letter used to indicate a population proportion, the proportion in a population possessing a particular attribute.
$\sqrt{\{\theta(1-\theta)/n\}}$	standard deviation of theoretical sampling distribution of proportions of samples, size n, when drawn from a population itself having proportion θ.
$z_i = \dfrac{x_i - \mu}{\sigma/\sqrt{n}}$	standard units value, or z value, of a sample mean x_i, in relation to the distribution of the sample means.
$z_i = \dfrac{p_i - \theta}{\sqrt{\{\theta(1-\theta)/n\}}}$	standard units value, or z value, of a sample proportion p_i, in relation to the distribution of the sample proportions.

Chapter 4

$\bar{x}$ (pronounced x-bar)	mean of a single sample.
s	standard deviation of a single sample.
$s/\sqrt{n}$	standard deviation of the theoretical sampling distribution of the mean, when σ is unknown and s is used as a 'best estimate' of σ.
standard error	standard deviation of a sampling distribution.
$t = \dfrac{\bar{x} - \mu}{s/\sqrt{n}}$	t value, comparable to a z value, where s is being used as a 'best estimate' of σ, and the

	sample size is less than 30. In these circumstances, the t statistic approximates to the Student-t distribution and not to the normal distribution.
Σ (pronounced sigma)	Greek (capital) letter used to indicate a summation of items. With suitable limits, the range of values summed is indicated.
$\displaystyle\sum_{i=1}^{n} (x_i - \bar{x})$	$= (x_1 - \bar{x}) + (x_2 - \bar{x}) + (x_3 - \bar{x}) + \ldots + (x_n - \bar{x}).$
$\sqrt{\{p(1 - p)/n\}}$	standard error of the proportion, when θ is unknown and p is used as the 'best estimate' of θ.
k	standard units value (z value) corresponding to a required probability.
N	population size.
$\sqrt{\left(\dfrac{N - n}{N - 1}\right)}$	small population correction. Factor by which the standard error is multiplied (and thereby reduced) when a sample is a substantial fraction of the population from which it is drawn.
$\mathrm{se}_{\bar{x}}$	standard error of the mean.
se_p	standard error of the proportion.

Chapter 5

$\bar{x}_1$, $\bar{x}_2$ (pronounced x one bar, x two bar)	means of different samples, distinguished by subscript.
p_1, p_2	proportions in different samples.
$z = \dfrac{\bar{x} - \mu}{s/\sqrt{n}}$	standard units value (large sample) for difference between sample and population means.
$z_{0.05}$	z value corresponding to 0.05 probability, that is, it excludes 0.05 (5%) of the total area under a standard normal curve, outside the values $-z_{0.05}$ to $+z_{0.05}$.
$z_{0.01}$, $z_{0.001}$	other subscripts refer similarly to other probabilities.
$\lvert\ \rvert$	indicates that the sign ($+$ or $-$) of the

	number or symbol contained between the upright parallel lines is to be ignored.
$\lvert z \rvert$	arithmetic value of z, irrespective of sign.
$\geqslant$	greater than, or equal in value, to the expression following.
$>$	greater than
$<$	less than
$\leqslant$	less than or equal to
μ'	population mean, the suffix distinguishing it from another population with mean μ.
g	difference between two population means in standard units.
μ_1, μ_2	means of populations, from which separate samples, size n_1 and n_2 (with means $\bar{x}_1$ and $\bar{x}_2$ respectively) are drawn.
σ_1, σ_2	standard deviations of populations with means μ_1 and μ_2.
s_1, s_2	standard deviations of samples with means $\bar{x}_1$ and $\bar{x}_2$.
$\mathrm{se}_{\bar{x}_1 - \bar{x}_2}$	the standard error of the difference between sample means, $\bar{x}_1 - \bar{x}_2$.

$$\mathrm{se}_{\bar{x}_1 - \bar{x}_2} = \sqrt{\left\{ \frac{\sigma_1^2}{n_1} + \frac{\sigma_2^2}{n_2} \right\}}$$ if the population standard deviations are known, or $\sqrt{\left\{ \dfrac{s_1^2}{n_1} + \dfrac{s_2^2}{n_2} \right\}}$ if the sample standard deviations are being used as 'best estimates' of the corresponding population standard deviations.

$$z = (\bar{x}_1 - \bar{x}_2) \Big/ \sqrt{\left\{ \frac{s_1^2}{n_1} + \frac{s_2^2}{n_2} \right\}}$$ standard units value for the difference between two sample means (large samples).

$$t = (\bar{x}_1 - \bar{x}_2) \Big/ \sqrt{\left\{ \frac{(n_1 - 1)s_1^2 + (n_2 - 1)s_2^2}{(n_1 - 1) + (n_2 - 1)} \left(\frac{1}{n_1} + \frac{1}{n_2} \right) \right\}}$$ t statistic for the difference between two sample means (small samples).

θ_1, θ_2	proportions in two separate populations.
p_1, p_2	proportions in single samples drawn from each of two populations.
$\mathrm{se}_{p_1 - p_2}$	standard error of the difference between two sample proportions, $p_1 - p_2$.

$se_{p_1 - p_2}$

$= \sqrt{\left\{ \dfrac{\theta_1(1 - \theta_1)}{n_1} + \dfrac{\theta_2(1 - \theta_2)}{n_2} \right\}}$, where the samples are drawn from two *separate* populations whose proportions are *known*, or

$\sqrt{\left\{ \dfrac{p_1(1 - p_1)}{n_1} + \dfrac{p_2(1 - p_2)}{n_2} \right\}}$, where the sample proportions are being used as 'best estimates' of the *unknown* population proportions θ_1 and θ_2, or

$\sqrt{\left\{ \theta(1 - \theta) \left(\dfrac{1}{n_1} + \dfrac{1}{n_2} \right) \right\}}$, where the samples are drawn from the *same* population whose proportion is *known*, or

$\sqrt{\left\{ p(1 - p) \left(\dfrac{1}{n_1} + \dfrac{1}{n_2} \right) \right\}}$ where the samples are drawn from the *same* population whose proportion is *unknown*, so that p is the 'best estimate' of θ, where

$$p = \frac{n_1 p_1 + n_2 p_2}{n_1 + n_2}.$$

z

$= \dfrac{p_1 - p_2}{se_{p_1 - p_2}}$, standard units value for difference between two sample proportions, with appropriate form of $se_{p_1 - p_2}$ (see above).

Chapter 6

I

represents the difference between two population parameters in situations which involve estimating the probable difference between population parameters on the basis of sample results.

$I_{0.95}$

the 0·95 probable difference between two population parameters.

Chapter 7

χ (pronounced ki)

Greek letter used to represent a multinomial distribution (see below).

o_i

general representation of *observed* value.

e_i

expected value, corresponding to *observed* value, o_i.

B

χ^2 (chi-square) $\qquad = \sum_{i=1}^{k} \frac{(o_i - e_i)^2}{e_i}$

r — number of rows in a contingency table.

k — number of columns in a contingency table.

$(r-1) \times (k-1)$ — number of degrees of freedom in an r by k contingency table.

C — contingency coefficient.

$C \qquad = \sqrt{\left(\frac{\chi^2}{\chi^2 + N}\right)}$

Introduction

1.1 From descriptive statistics to statistical methods

As a term, the word 'statistics' tends to be used very loosely, referring to anything connected with numerical data. At one extreme is the mass of original data: census material; records of central and local government, industry and commerce; and data collected in sample surveys and other inquiries. From these can be derived descriptive statistics of various kinds. These range from the type of 'head counting' implicit in the summaries of data contained in frequency distributions, and other forms of tabulation, to central measures and measures of dispersion or variation. The latter are specific to the data from which they are obtained. A particular set of data has for example, one mean, one median, one standard deviation; and each of these measures is unique for the data from which it is derived. The reverse is not true, for a particular mean and standard deviation are not generated by one, and only one, set of data. A little thought and experimentation with actual numbers soon establishes this. But the unique nature of these descriptive measures, and the mathematical relationship which some of them have with their basic data (notably the mean and standard deviation), makes them of great value in the field of summarizing and interpreting numerical material.

Descriptive measures end at description. They do not go beyond the point of summarizing, providing for interpretation through the calculated measures and permitting comparisons between two or more sets of data. They are basically factual. They do not extend into the field of statistical methods, which in the broadest sense is concerned with deduction and inference. Statistical methods aim at going beyond the descriptive, factual field into, for example, the realm of estimation, hypothesizing, projection into the future and the building of models. All these aspects of statistical methods have one thing in common; they aim at the production of information about statistical relationships and further, they may also provide a basis for decision-making. The results are highly informative in a specific way.

This book is concerned with one broad aspect of statistical methods, that which involves the relationship between samples and the populations from which they are drawn. The field is looked at first as the relationship between a single sample and its parent population, and secondly in a wider sense where more than one sample and more than one population are involved, and finally in situations involving sets of sample data.

1.2 Inference from sample results

The use of sampling as a process has developed almost entirely in the last half century. Before 1939, fairly limited use was made of sampling procedures in the UK in economic and social fields. In the USA, during the same period, the use of sampling had extended further. Statistical surveys began in the UK with B. S. Rowntree's survey of York, which took place in 1899, and is described in *Poverty: A Study of Town Life* (1902). This was the first attempt to carry out a rigorous survey and to measure the extent of poverty according to a closely drawn definition. What Rowntree did was to carry out a survey of all working-class households in York, thus providing himself with a complete population, rather than a sample. At that time, survey sampling was an unknown art. When he carried out a further survey of York in 1936, Rowntree used the same method of complete enumeration of working-class households. However, in a most interesting supplementary chapter at the end of his book *Poverty and Progress* (1941), which he addressed not to the general reader, but to statisticians and others who might make social surveys of other cities, he examined the reliability of social statistics based on the sampling method. This he did by studying the samples which could be obtained by taking every 10th, 20th, 30th, 40th and 50th of his schedules, and calculating from each of these samples various statistics which he had previously calculated for the population. After examining these results, Rowntree's conclusions were that for measures based on relatively large groups, a sample of 1 in 10 would provide sufficiently accurate figures. His last survey of York, carried out in 1950 in conjunction with G. S. Lavers, and described in *Poverty and the Welfare State* (1951) was based on a 1 in 9 (approximately 11 per cent) sample of working-class households.

The expansion of sampling and surveys using sampling methods was part of the great post-1945 expansion in the collection and use of statistical material generally. This expansion is still continuing. The last twenty years has seen a vast increase in the number of surveys and other inquiries making use of some form of sampling procedure, and with this has come an increased sophistication in both the actual techniques used in obtaining sample data, and in the inferential processes to which the data have been subjected. Students of economics, sociology, business studies, accountancy and related fields will inevitably find themselves studying material which includes surveys or other sample inquiries. It is becoming increasingly difficult to do this without having a sufficient knowledge of the statistical techniques involved in order to appraise critically the methodology used, and the conclusions reached as a result.

Any inquiry must serve a purpose. A sample survey or other inquiry based on sampling techniques is aimed at discovering something about the population from which the sample is drawn. Population here simply means the whole group of items with which the survey is concerned and from which the sample is taken. It may consist of people in general, for example, the whole population of Great Britain; or people with specific characteristics, such as men between twenty-one and forty-five years of age. Alternatively, the population may consist of objects or institutions, rather than individual people, for example, items taken from a factory production line, or local education authority primary schools. The sample is taken to represent the population. How far can this be true? Sampling is here taken to be random, that is, the actual choice of items to be included in the sample has been left to chance. No bias, no human interference, is allowed. A sample thus obtained should be reasonably representative of the population. But how representative? To what extent might the sample be unrepresentative of the population, even though the sample is unbiased and has been chosen by random means? A sample might nonetheless consist largely of extreme items. No one expects a random sample to represent *exactly* the population from which it is drawn. The results of tossing a balanced coin 100 times can be looked upon as a random sample drawn from an infinitely large population consisting theoretically of equal numbers of heads and tails. The population parameter, here the proportion of heads, would be one half. It is possible that the sample statistic, the proportion of heads in the sample, might also be one half. But if the sample arising from 100 tosses of a balanced coin consisted of, say, 48 heads and 52 tails, this would hardly lead to the conclusion that the coin was not balanced, or that there was some bias in the tossing. Opinions would be different if only 30 heads appeared against 70 tails. That sort of result would give rise to the strong suspicion, either that the coin was unbalanced or that selection was non-random.

Statistical inference helps to answer these questions. It expresses limitations on the use of sample statistics as equivalent to population parameters. It provides a cautionary reminder that there is no certainty in the conclusions to be drawn from inquiries using sampling; no certainty—only degrees of probability. To investigate statistical inference therefore, it is necessary to start with a study of probability.

1.3 Probability

The basis of statistical inference is probability, the probability that the sample will reflect in certain ways the population from which it is drawn. It is the application of the basic rules of probability which enables any statistical inference to be made.*

In mathematics generally, total probability is looked upon as being equal to 1. Individual probabilities are looked upon as fractions or decimal fractions of 1. Consider the tossing of a coin. The possible results are that the coin appears with the head uppermost, or with the tail uppermost.† These are mutually exclusive events; if one happens the other cannot, and vice versa. Together they exhaust probability—there is nothing else that can happen. If, therefore, p_1 is the probability of throwing a head, and p_2 is the probability of throwing a tail,

$$p_1 + p_2 = 1.$$

This result can be expressed in another way:

$$p_1 = 1 - p_2$$

that is, the probability of throwing a head is equal to one minus the probability of throwing a tail. It can be verbally, and more usefully, expressed the reverse way round as well; the probability of *not* throwing a head is equal to one minus the probability of throwing a head.

Although it may seem a tortuous way of stating the obvious, this particular way of expressing probability can be very useful. In general terms, if the probability of something happening is p, then the probability of it *not* happening is $(1 - p)$.

Suppose that a dice is thrown. It has six faces, one of these,

---

* Probability is dealt with in G. R. Wyborn, *Elementary Mathematics for Applied Statistics*, but for convenience, the basic rules are repeated here.

† I am deliberately ignoring the small and irrelevant probability that the coin might land on its edge.

and only one of them, will be the uppermost face, showing a 1, 2,... or 6. These results exhaust probability, and are mutually exclusive, in the sense that only one can occur at a time. If, therefore, p_1 is the probability of throwing a 1, etc.

$$p_1 + p_2 + p_3 + p_4 + p_5 + p_6 = 1$$

If the dice is balanced, the six possible results will be equiprobable, so that

$$p_1 = p_2 = p_3 = p_4 = p_5 = p_6 = 1/6.$$

The probability of throwing a six is $p_6 = 1/6$. The probability of *not* throwing a six is $(1 - p_6) = 5/6$. This is the same as the probability of throwing a 1, 2, 3, 4 or 5, and is equal to the sum of their separate probabilities.

Whether tossing a coin or throwing a dice, only *one* result can be obtained. For this reason the throwing of a head or a tail are both known as mutually exclusive events, as has been referred to above. Similarly, the rolling of a 1 or 2 etc. with a dice are again mutually exclusive events. If a head is thrown, a tail cannot be thrown; if a 4 is rolled with a dice, no other number can be uppermost.

So far this can be summarized by saying:

where a, b,..., k are mutually exclusive events which together exhaust probability, and p_a is the probability of event a occurring

$$p_a + p_b + \ldots + p_k = 1. \qquad \text{RULE 1}$$

This can be further expressed as:

$$p_a = 1 - (p_b + p_c + p_d + \ldots + p_k)$$

which is the probability of event *a* occurring. The probability of event *a* not occurring is:

$$(1 - p_a) = (p_b + p_c + \ldots + p_k).$$

Suppose that a dice is thrown, and that a five or six is wanted. The throwing of a five and the throwing of a six are mutually exclusive events. The probability that one of these numbers will appear is the sum of their individual probabilities, here $(1/6 + 1/6)$ or $1/3$. In general terms, and using the previous notation, the probability that one of two mutually exclusive events, *a* and *b*, will occur is $(p_a + p_b)$. This can be extended to cover any number of mutually exclusive events:

where a, b, c, ..., k are a number of mutually exclusive

events, the probability that one of them will occur is given by the sum of their separate probabilities,

$$(p_a + p_b + \ldots + p_k).$$ RULE 2

So far, a single event or result only has been considered. Suppose, however, that a dice is thrown twice. The probability of getting a six the first time is 1/6; if the dice is thrown a second time, the probability of getting a six is again 1/6. These are independent events. The result on the second throw is in no way influenced by the result of the first.* The probability of getting a six in two successive throws is the probability of getting a six on the first throw, multiplied by the probability of getting a six again on the second throw. It is, therefore, 1/6 × 1/6 or 1/36. This can be expressed in general terms by saying that where two independent events, *a* and *b*, are both to occur, the probability of them so doing is $(p_a \times p_b)$. For any number of such events:

where a, b, c, ..., k are independent events, the probability of them all occurring, either simultaneously or consecutively,† is the product of their separate probabilities,

that is, $(p_a \times p_b \times \ldots \times p_k).$ RULE 3

So far discussion has centred on mutually exclusive events, but this is not always the case. Suppose that the two events are not mutually exclusive. If a dice is thrown twice, a six might appear on the first throw, or on the second, or on both. The appearance of a six on the first throw does not prevent a six appearing on the second. Where events are not mutually exclusive, the probability of at least one of the events *a* and *b* occurring has to take account of the probability that both may occur. So that in terms of throwing a dice twice, the probability of getting at least one six has to take account of the probability of getting two sixes. The general statement is:

---

* The independence of events of this kind is sometimes a little difficult to accept. If *on the average* one-sixth of all dice thrown should produce a six, then the appearance of a six on one throw should make it less likely to appear on the next, so the argument runs. In the long run, the number of sixes thrown will approximate to one-sixth of the throws. The emphasis, however, is on 'the long run', which is concerned with large numbers. In the short run, no such 'average' applies, and the result of the second throw of the dice is in no way influenced by the outcome of the first. It is important to realize that dice do not have memories.

† I mention here events occurring 'either simultaneously or consecutively'. The logic in dealing with both situations is the same, and simultaneous events, e.g. two dice rolled together, are often more easily dealt with in probability terms as consecutive events, i.e. one dice thrown twice successively.

where events are not mutually exclusive, the probability of at least one of the events a and b occurring is

$$[p_a + p_b - (p_a \times p_b)].$$ RULE 4

Reverting to the dice throwing, the probability of throwing at least one six in two throws is $[p_6 + p_6 - (p_6 \times p_6)] = 1/6 + 1/6 - (1/6 \times 1/6) = 1/3 - 1/36 = 11/36$.

This can easily be checked by considering the probability of not throwing a six either time since throwing a six at least once, and not throwing a six at all, together exhaust probability. The probability for not throwing a six is $(1 - p_6)$, or $(1 - 1/6)$, that is, 5/6. Two successive throws represents two independent events, so that the probability of not throwing a six on both occasions is $(1 - p_6) \times (1 - p_6)$ or $5/6 \times 5/6 = 25/36$. The probability of getting at least one six in two throws must be $(1 - 25/36)$ or 11/36. This confirms the result obtained by the direct approach.

Tree diagrams

It is often easier to establish probabilities by considering, step by step, all possible sequences of events, and then assigning to the sequences their appropriate probabilities.

Consider the rolling of a dice three times, and the way in which sixes might appear. Probabilities are shown in brackets in Diagram 1.1.

Diagram 1.1 *Rolling dice: probability of sixes appearing*

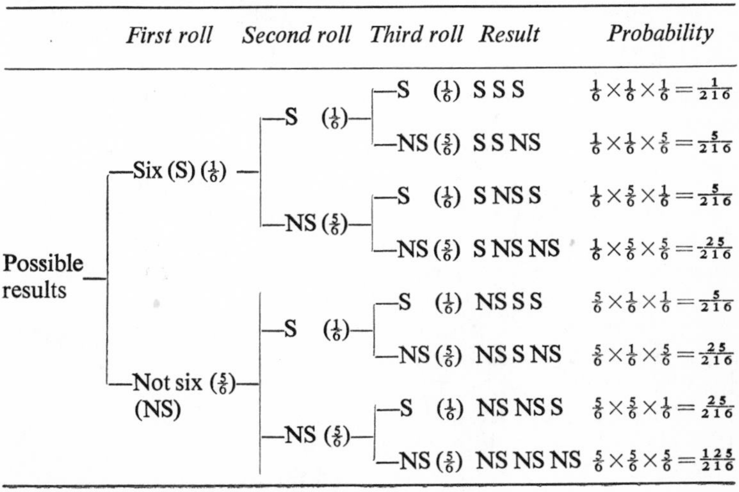

	First roll	Second roll	Third roll	Result	Probability
Possible results	Six (S) $(\frac{1}{6})$	S $(\frac{1}{6})$	S $(\frac{1}{6})$	S S S	$\frac{1}{6} \times \frac{1}{6} \times \frac{1}{6} = \frac{1}{216}$
			NS $(\frac{5}{6})$	S S NS	$\frac{1}{6} \times \frac{1}{6} \times \frac{5}{6} = \frac{5}{216}$
		NS $(\frac{5}{6})$	S $(\frac{1}{6})$	S NS S	$\frac{1}{6} \times \frac{5}{6} \times \frac{1}{6} = \frac{5}{216}$
			NS $(\frac{5}{6})$	S NS NS	$\frac{1}{6} \times \frac{5}{6} \times \frac{5}{6} = \frac{25}{216}$
	Not six $(\frac{5}{6})$ (NS)	S $(\frac{1}{6})$	S $(\frac{1}{6})$	NS S S	$\frac{5}{6} \times \frac{1}{6} \times \frac{1}{6} = \frac{5}{216}$
			NS $(\frac{5}{6})$	NS S NS	$\frac{5}{6} \times \frac{1}{6} \times \frac{5}{6} = \frac{25}{216}$
		NS $(\frac{5}{6})$	S $(\frac{1}{6})$	NS NS S	$\frac{5}{6} \times \frac{5}{6} \times \frac{1}{6} = \frac{25}{216}$
			NS $(\frac{5}{6})$	NS NS NS	$\frac{5}{6} \times \frac{5}{6} \times \frac{5}{6} = \frac{125}{216}$

Each combined result is unique, because of the ordering of the rolls of the dice. At one extreme, the probability of getting three sixes is 1/216, a combination which occurs once only. The probability of getting no sixes at all is 125/216, and of getting one or more sixes 1 − 125/216, or 91/216.

However, if the order of the result is not relevant, but only the combination ultimately appearing, then some combinations are identical. Thus, whereas there is only one result for three sixes, and only one for no sixes, there are three combinations showing one six only, and three containing two sixes. Disregarding order then, the probability of one six occurring is 3 × 25/216 or 75/216 (the number of different ordered results multiplied by the probability for the result) and for two sixes the probability is 3 × 5/216 or 15/216. Total probability is still 1. The situation in which order is ignored is important, for this conforms to the case of simultaneous as opposed to consecutive events. In this example above, the results are then those that would obtain from rolling three dice simultaneously instead of successively. Although what may be required are the results of simultaneous events, in practice it may be easier to analyse the situation by considering the events to be consecutive, and then adjusting the conclusions for the events being simultaneous.

This now provides the basis for using probability in the development of sampling theory. However, the reader may care to try some of the following exercises involving the practical operation of simple probability theory.

Exercises

For questions 1–4 assume a well-shuffled standard pack of 52 playing cards, and that selection is at random.

1. What is the probability of drawing (*a*) a red card (*b*) a spade (*c*) a queen (*d*) a red five?

2. What is the probability of drawing (*a*) an even-numbered card (excluding royalty) (*b*) a king or an ace (*c*) a spade or a ten?

3. If two successive cards are drawn (with replacement) what is the probability of drawing (*a*) two black cards (*b*) two diamonds (*c*) two royal cards (from Jack, Queen and King) (*d*) a red card and a black card?

4. Repeat 3. above, but without replacement between draws.

5. Mr A may walk home from the station, take a bus or ring his wife to fetch him by car. If the probability of Mr A walking home is 0·3, and of his taking a bus 0·6, what is the probability that his wife fetches him home by car?

6. (*a*) If the probability is 0·4 that the canteen will serve

sausages for lunch and $0 \cdot 3$ that it will serve apple pie, what is the probability that on any one day it will serve both sausages and apple pie?

(*b*) What is the probability that the canteen will be serving sausages, or apple pie, or both? (i.e. at least one of these dishes).

7. If the probability of rain on any day in June is $0 \cdot 3$, and the probability that Mr B (a keen swimmer), will go swimming is $0 \cdot 2$ if it is raining but $0 \cdot 8$ if it is fine, what is the probability of Mr B swimming on a day in June?

The following two questions involve practical experiments; tree diagrams help a lot.

8. Toss six coins 128 times and for each tossing, record the number of heads that appear. (This is done more easily if six people each toss one coin.) Work out the probability for each possible result. Then by multiplying together the probability and total number of tosses, obtain the theoretical distribution. This you can compare with the distribution you have constructed by practical experiment. Provided there was no bias in the tossing, the two distributions should be reasonably close. You can test this statistically later on (see section 7.4).

9. Take eight playing cards, of which two are aces. Shuffle the cards well, so that they are in random order. Take it in turns with a friend to draw a card (without replacing it) until one of you draws an ace. Record who wins—*A* (who has first draw) or *B* (who has second draw). Repeat the shuffling and drawing for a number of times, say 50, but more if you can. What is the probability of *A* winning? What are the advantages of the first draw?

Then work out the theoretical probabilities. Don't forget— the probability of *B* getting a turn is the same as the probability that *A* does not draw an ace, and so on for *A*'s second turn, *B*'s second turn, etc.

Further reading

A light-hearted but instructive introduction to the laws of probability is to be found in:

D. Huff, *How to Take a Chance* (Penguin Books, 1965).

Informal discussions of probability are to be found in:

J. E. Freund and F. J. Williams, *Modern Business Statistics* (Pitman, 2nd ed. 1970), chapter 6.

M. J. Moroney, *Facts from Figures* (Penguin Books, 1951), chapter 2.

W. J. Reichmann, *Use and Abuse of Statistics* (Penguin Books, 1964), chapter 14.

W. A. Wallis and H. V. Roberts, *Statistics: A New Approach* (Methuen, 1957), chapter 10.

Basic distributions

2.1 The binomial distribution

In chapter 1, when discussing probability, the results of rolling a dice or tossing a balanced coin a number of times were considered. The results obtained were in fact examples of the binomial distribution which now needs to be examined further.

Consider the tossing of a balanced coin one, two, three and four times, using a tree diagram (see Diagram 2.1).

This can now be summarized, *ignoring order*, to obtain the results of simultaneous throws:

First toss	Second Toss	Third toss	Fourth toss
1 (H)	1 (2H)	1 (3H)	1 (4H)
1 (T)	2 (HT)	3 (2HT)	4 (3HT)
	1 (2T)	3 (H2T)	6 (2H2T)
		1 (3T)	4 (H3T)
			1 (4T)

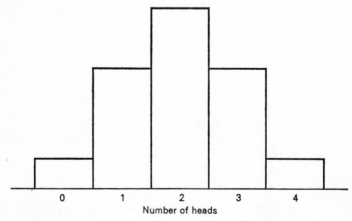

Number of heads

Figure 2.1

Diagram 2.1   *Tossing a balanced coin*

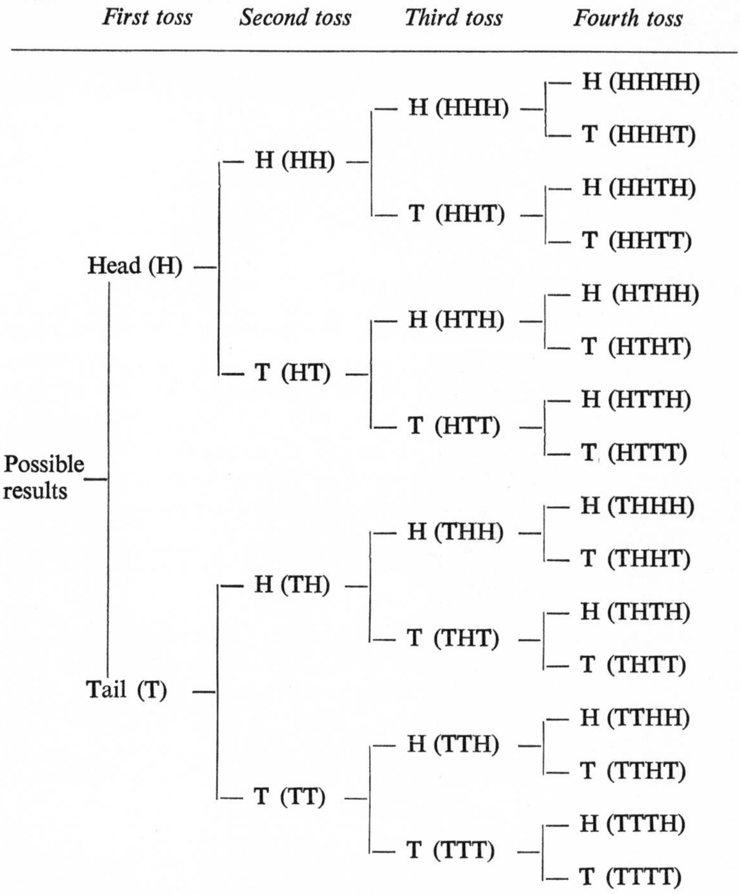

First toss	*Second toss*	*Third toss*	*Fourth toss*

Possible results

Head (H)
- H (HH)
  - H (HHH)
    - H (HHHH)
    - T (HHHT)
  - T (HHT)
    - H (HHTH)
    - T (HHTT)
- T (HT)
  - H (HTH)
    - H (HTHH)
    - T (HTHT)
  - T (HTT)
    - H (HTTH)
    - T (HTTT)

Tail (T)
- H (TH)
  - H (THH)
    - H (THHH)
    - T (THHT)
  - T (THT)
    - H (THTH)
    - T (THTT)
- T (TT)
  - H (TTH)
    - H (TTHH)
    - T (TTHT)
  - T (TTT)
    - H (TTTH)
    - T (TTTT)

The histogram in Figure 2.1 depicts the results.

The number of times a particular combination occurs (but each time in a different order) indicates the relative frequency (or probability) of the combination. As the coin is balanced, a head and a tail are equiprobable ($p = 1/2$ or $0\cdot5$), and the probabilities of the results are, therefore those given in Table 2.1.

In each case total probability is one, and the distributions of the different sets of results are relative frequency distributions, or probability distributions. These probabilities arise from the expansion of the algebraic distribution $(a + b)^n$, which is known as the binomial distribution, for the special case where

Table 2.1 *Tossing a balanced coin—probabilities*

First toss	Second toss	Third toss	Fourth toss
$p\,(\mathrm{H}) = \frac{1}{2}$	$p\,(2\mathrm{H}) = \frac{1}{2} \times \frac{1}{2} = \frac{1}{4}$	$p\,(3\mathrm{H}) = (\frac{1}{2})^3 = \frac{1}{8}$	$p\,(4\mathrm{H}) = (\frac{1}{2})^4 = \frac{1}{16}$
$p\,(\mathrm{T}) = \frac{1}{2}$	$p\,(\mathrm{HT}) = 2 \times (\frac{1}{2})^2 = \frac{1}{2}$	$p\,(2\mathrm{HT}) = 3 \times (\frac{1}{2})^3 = \frac{3}{8}$	$p\,(3\mathrm{HT}) = 4 \times (\frac{1}{2})^4 = \frac{4}{16}$
	$p\,(2\mathrm{T}) = \frac{1}{2} \times \frac{1}{2} = \frac{1}{4}$	$p\,(\mathrm{H}2\mathrm{T}) = 3 \times (\frac{1}{2})^3 = \frac{3}{8}$	$p\,(2\mathrm{H}2\mathrm{T}) = 6 \times (\frac{1}{2})^4 = \frac{6}{16}$
		$p\,(3\mathrm{T}) = (\frac{1}{2})^3 = \frac{1}{8}$	$p\,(\mathrm{H}3\mathrm{T}) = 4 \times (\frac{1}{2})^4 = \frac{4}{16}$
			$p\,(4\mathrm{T}) = (\frac{1}{2})^4 = \frac{1}{16}$

$(a + b) = 1$. In this example, a and b are respectively the probability of obtaining a head and the probability of obtaining a tail, and n, the number of coins tossed. In the above example, $a = 1/2$, $b = 1/2$, and n takes the values from 1 to 4. But, a and b do not have to be equal. As explained earlier, if a dice is rolled for sixes, $a = 1/6$ (probability of getting a six) and $b = 5/6$ (probability of not getting a six). The pattern of the different combinations is the same. However, the equivalent relative frequencies or probabilities for one, two, three, and four rolls would be those given in Table 2.2.

The results are shown in the histogram in Figure 2.2.

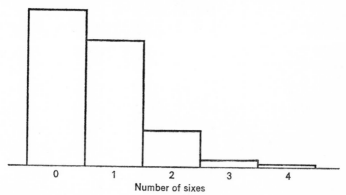

Number of sixes

Figure 2.2

The general form of the binomial distribution as used here is $[p + (1 - p)]^n$ where, as explained in chapter 1, p is the probability of an event occurring, $(1 - p)$ of it not occurring and n the number of 'trials'.

So far, n has been taken through values of 1 to 4 with $p = 1/2$ and $p = 1/6$. But it can take on any value, that is represent any number of tosses, throws, or more generally 'trials'. The diagrams in Figure 2.3 are examples of distributions with different values of n.

General terms can be developed from the binomial distribution for any arithmetical values of p and n.

Suppose the probability of success in a single 'trial' is p, and that n trials are conducted; the *possible* number of 'successes' is $0, 1, 2, \ldots, n$. From the binomial distribution can be obtained the *probability* of getting $0, 1, 2, \ldots, n$ successes. The probability of getting r successes in n trials is

$$\binom{n}{r} \times p^r \times (1 - p)^{n-r}$$

Table 2.2   *Rolling dice for sixes—probabilities*

First roll	Second roll	Third roll	Fourth roll
$p\,(\text{Six}) = \frac{1}{6}$	$p\,(2\text{S}) = (\frac{1}{6})^2 = \frac{1}{36}$	$p\,(3\text{S}) = (\frac{1}{6})^3 = \frac{1}{216}$	$p\,(4\text{S}) = (\frac{1}{6})^4 = \frac{1}{1296}$
$p\,(\text{Not six}) = \frac{5}{6}$	$p\,(\text{SN}) = 2 \times \frac{1}{6} \times \frac{5}{6} = \frac{10}{36}$	$p\,(2\text{SN}) = 3 \times (\frac{1}{6})^2 \times (\frac{5}{6}) = \frac{15}{216}$	$p\,(3\text{SN}) = 4 \times (\frac{1}{6})^3 \times (\frac{5}{6}) = \frac{20}{1296}$
	$p\,(2\text{N}) = (\frac{5}{6})^2 = \frac{25}{36}$	$p\,(\text{S2N}) = 3 \times \frac{1}{6} \times (\frac{5}{6})^2 = \frac{75}{216}$	$p\,(2\text{S2N}) = 6 \times (\frac{1}{6})^2 \times (\frac{5}{6})^2 = \frac{150}{1296}$
		$p\,(3\text{N}) = (\frac{5}{6})^3 = \frac{125}{216}$	$p\,(\text{S3N}) = 4 \times \frac{1}{6} \times (\frac{5}{6})^3 = \frac{500}{1296}$
			$p\,(4\text{N}) = (\frac{5}{6})^4 = \frac{625}{1296}$

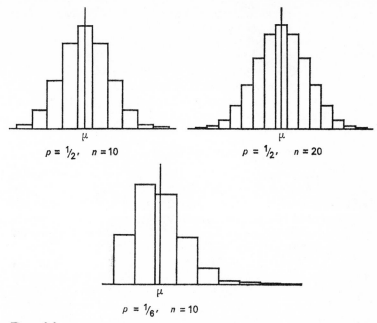

Figure 2.3

where $\binom{n}{r}$ is the appropriate binomial coefficient,* and r takes values from 0 to n (as above). To revert to previous data about rolling dice, where $n = 4$ and $r = 2$, $\binom{n}{r} = 6$. The probability of getting two successes in n trials where the probability of success is 1/6 (and of non-success 5/6) is therefore:

$$6 \times \left(\frac{1}{6}\right)^2 \times \left(\frac{5}{6}\right)^2 \text{ or } \frac{150}{1296}$$

which is the same as obtained previously.

* The binomial coefficients are obtained from the expansion of $(a + b)^n$ for the special case where $a = b = 1$. There is a different set of coefficients for each value of n. The calculation of binomial coefficients is not difficult (see below) but a number of textbooks on algebra or statistics contain tables of binomial coefficients for values of n up to about 20.

The coefficient $\binom{n}{r}$ relates to the term r successes in n trials, and is the number of times a particular set of results can be expected, disregarding the order of results. Algebraically $\binom{n}{r} = \dfrac{n!}{r!(n-r)!}$, where $n!$ (referred to as 'n-factorial') $= n \times (n - 1) \times (n - 2) \times \ldots \times 3 \times 2 \times 1$. (For example, if $n = 7$, then $n! = 7 \times 6 \times 5 \times 4 \times 3 \times 2 \times 1$.)

C

It will not be necessary for the majority of students using this book to develop the theoretical aspects of the binomial distribution. What is necessary is to understand that probability data of a particular kind, involving a series of 'trials' in which the result is either a 'success' or a 'failure', follow the pattern of the binomial distribution. The properties of the binomial distribution are therefore of great importance in understanding the practical operation of probability.

2.2   The normal distribution

The binomial distribution is a discrete (or discontinuous) distribution. That is, it takes on specific values only. If coins are tossed, then 0, 1, 2, etc., heads may appear, but there can be no value between these, such as $1 \cdot 2$ heads or $2 \cdot 9$ heads. In actual surveys or other forms of data collection, this is often a reasonable representation of a real situation. For example, in an opinion poll type of survey, the number of persons subscribing to a particular viewpoint can take on whole-number values only. However, in many ways the binomial distribution is awkward to use. In order to calculate the probability of getting say three or fewer heads when a coin is tossed ten times, it is necessary to calculate separately the probability of getting 3, 2, 1 and 0 heads. Although this may be tolerable when dealing with small numbers, it becomes quite impossible in dealing with larger numbers. It would be very tedious to work out the probability distribution for the result of asking 100 people which party they would vote for in a general election if one were held tomorrow; yet this, and far more complicated situations, arise in real life.

Further, some data are continuous, that is, all possible values can occur. Age is an example, height and weight are other examples. Although age may be measured in specific units such as months, days, or years, to two places of decimals, it actually takes a continuous form. To reach his present age a person has passed through all possible ages which are less than his present age. Mathematically speaking, continuous distributions are easier to deal with than discrete distributions. The treatment of frequency distributions for the calculation of mean and standard deviation is based on the distributions being continuous, whether or not the basic data is itself continuous.

Calculated statistics, such as the mean and standard deviation, would behave as continuous data irrespective of the form (discrete or continuous) which is taken by the basic material. Suppose ten coins are tossed 100 times. The number of heads appearing at each trial will be a whole number varying from 0 to 10. A frequency distribution derived from the results would be discrete. But if the mean number of heads per trial is calculated from the frequency distribution, this statistic could take on far more than eleven values, since it would not be limited to the whole-number values 0 to 10. The mean would in fact be a continuous variable.

Where continuous data are involved, another distribution, the *normal distribution*, is appropriate. Knowledge of this distribution dates back to the eighteenth century; it appears first to have been developed by Abraham de Moivre (1667–1745), who considered it had practical application only in relation to games of chance. Carl Gauss (1777–1855), however, used the curve as the basis of a theory about accidental errors of measurement. Empirically it was observed that measurements which should have been identical in fact differed slightly, and that the actual measurements obtained formed a pattern. This pattern was referred to as the 'normal curve of errors' since it was considered to represent the errors obtained in making repeated measurements. Suppose, for example, that a scientific experiment is carried out a large number of times under apparently identical conditions. The results will not necessarily be exactly the same each time, but variations should be very slight, and small variations will occur more often than large ones. The pattern would be that described by the 'normal curve of errors',* which is shown in Figure 2.4.

Although originally the normal curve was related to the patterns of observed error, in practice other data tend to approximate to this distribution. Height is an example of such a variable; but, in addition, a great deal of statistical material (here meaning simply collected numerical data) appears in frequency distributions which approximate to the bell-shaped normal distribution. Even though the approximation may not be very close, the properties of the normal distribution can

* This principle is much used in the industrial field, in quality control, in a wide variety of situations. Very slight variations in the finish, dimensions, tensile strength, etc., of a production item are inevitable, and are accepted as common to the industrial process. This variation occurs despite apparently identical conditions under which the production process takes place. The problem here is ensuring that such variations do not exceed acceptable limits.

Another way of looking at this in a very simple example is to consider a group of fifty people sent out with identical tape measures to measure a particular wall. It would be surprising indeed if the resulting fifty measurements were identical. What would be expected would be that the measurements were fairly close together within a narrow range, and that the majority were clustered round some central value, which was itself very close to the actual value of the wall measurement.

still be used. Thus the reference to errors has been dropped from its description, and it is referred to now simply as the normal curve or the normal distribution.

The normal distribution is of great importance in the field of statistical methods. Certain theoretical distributions which are the basis of sampling theory are normal distributions. Further,

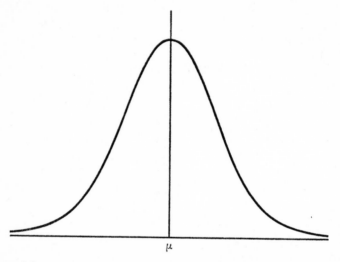

Figure 2.4

as a continuous distribution, the normal distribution can be used as a convenient substitute for the binomial distribution, as there are many occasions when there is a very close approximation between the two. The properties of the normal distribution, therefore, need to be studied and understood.*

The formula for the normal distribution in conventional co-ordinate geometry terms is:

$$y = \frac{1}{\sigma\sqrt{(2\pi)}} \, exp\left\{-\frac{1}{2}\left(\frac{x-\mu}{\sigma}\right)^2\right\}$$

where π is a constant, being the ratio of the circumference of a circle to its diameter. The exponential form, which can be written as e^x instead of $exp(x)$, is a power of the irrational constant e whose value to five places of decimals is $e =$

---

* It will be noted that the normal curve, shown in Figure 2.4, approaches, but does not actually meet the horizontal axis representing $y = 0$. Mathematically, the normal curve is referred to as being asymptotic to this axis.

2·71828.* If, therefore, for a given set of data the values of the mean and standard deviation, μ† and σ, are known, these can be substituted in the above formula, and the value of y (height of curve) can then be obtained for any given value of the variable x. All properties of the normal distribution can be deduced once the paired values of x and y are obtained, and these in their turn depend on the two parameters, the mean and standard deviation of the distribution.

Remembering the normal distribution formula is of little relevance, for it is not directly used in this form. The important point here is that for a given set of data, the normal distribution depends entirely on the two variable measures, the mean and the standard deviation. If, therefore, the mean and standard deviation are known for data which approximate to the normal distribution, all other mathematical properties of the distribution can be deduced. For practical purposes, this is very valuable.

2.3 The standard normal distribution

Because the normal distribution is so useful, a form has been derived which can be related to any set of data, irrespective of the actual mean and standard deviation as measured in the original units of the data. The properties of this form are embodied in what is commonly called a normal curve area table. The equation for the normal distribution referred to in the preceding section relates values of x and y, where y is the height of the curve corresponding to any given value of the variable x. This is not directly useful, since it is the area under a curve, (or histogram, or frequency polygon) which is related to frequency. The practical application of the normal distribution is concerned with the relative frequency (or probability) with which given values of x occur. That is, the area under any curve between two given values, x_1 and x_2, represents the frequency with which values between x_1 and x_2 occur. If this area is then expressed as a fraction of the total area, it will give the relative or proportionate frequency with which values between x_1 and x_2 occur. In its turn, relative frequency can represent the probability with which values occur between x_1 and x_2.

* e is the sum of the infinite expansion $\dfrac{1}{0!} + \dfrac{1}{1!} + \dfrac{1}{2!} + \dfrac{1}{3!} + \dfrac{1}{4!} + \dfrac{1}{5!} + \cdots$ and is the basis of the Napierian system of logarithms. And, for example, $exp\,(4) = e^4 = 2·718284$.

† The Greek letter μ (pronounced 'mew') is conventionally used in statistical texts to represent the population mean, and the Greek letter σ (pronounced 'sigma') to represent the population standard deviation.

The principle behind this common form is to convert all the measurements from the original data into what are known as standard units. These are often referred to as 'z values'. Standard units are obtained by converting the original measurements into deviations from the mean of their distribution $(x - \mu)$, and then measuring these deviations in units of the standard deviation of the distribution, i.e. dividing by σ. Instead of x_i, the measurement becomes

$$z_i = \frac{x_i - \mu}{\sigma}$$

μ and σ being the mean and standard deviation, which are of course constants for the given data. The effect of this operation is to convert the original distribution into a standard distribution in terms of $z_1, z_2, \ldots, z_n$, which has a mean of 0 and a standard deviation of 1. This eliminates from the data the effect of the order of magnitude of the original measurements and the extent of their variation. Figure 2.5 may help to illustrate this point.

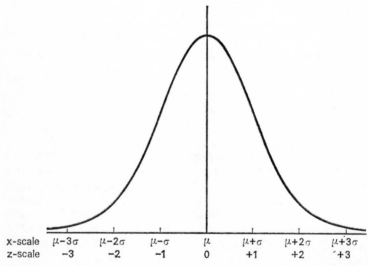

x-scale	$\mu - 3\sigma$	$\mu - 2\sigma$	$\mu - \sigma$	μ	$\mu + \sigma$	$\mu + 2\sigma$	$\mu + 3\sigma$
z-scale	−3	−2	−1	0	+1	+2	+3

Figure 2.5

The z values are ratios, so that the units of the original data also disappear. Further, the standard normal distribution is arranged so that total frequency is 1. That is, all actual frequencies becomes fractions of a total frequency of 1. This is analogous to the percentage frequency distribution. All normal distributions can be turned into this common form.

The normal curve area table relates z values to the area between the mean of the distribution (0) and the given z value (see Appendix, Table I). The table operates very much like a logarithm table, allowing for z values to two places of decimals up to $z = 5$. For the given z value the table gives the area from the mean of 0 to the positive z value. Since the curve is symmetrical, areas in the negative half of the curve can be obtained similarly. That is, the area from 0 to $-z$ is the same as that from 0 to $+z$.*

The following examples show the mechanics of how other areas, and thus the proportion of frequency falling between any two z values, can be obtained. In each case a diagram helps to clarify the process, with the area to be obtained shown hatched.

1. Find the area between $z_1 = 1 \cdot 05$ and $z_2 = 2 \cdot 3$.

The difference between the two areas is required (see Figure 2.6).

Area from 0 to $z_2 = 2 \cdot 3$ is $0 \cdot 4893$.
Area from 0 to $z_1 = 1 \cdot 05$ is $0 \cdot 3531$.
Therefore area from $z_1 = 1 \cdot 05$ to $z_2 = 2 \cdot 3$ is $0 \cdot 1362$.

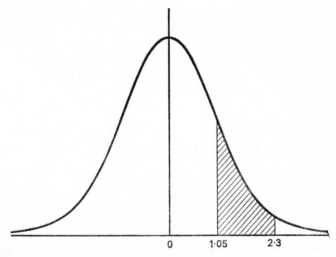

Figure 2.6

* Mathematically the normal distribution is an open-ended distribution, that is, it can take on infinitely high or low values. But these extreme values occur with infinitely small frequencies, as is shown by looking at the normal curve area table. The area under half the normal curve reached for say $z = 5$, is clearly very

2. Find the area between $z_1 = -0.96$ and $z_2 = 1.44$.
   Here the areas must be added together (see Figure 2.7).
   Area from 0 to $z_1 = -0.96$ is 0.3315.
   Area from 0 to $z_2 = 1.44$ is 0.4251.
   Therefore area from z_1 to z_2 is 0.7566.

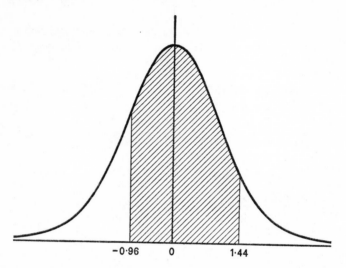

Figure 2.7

3. Find the area to the right of $z = 0.63$ (see Figure 2.8).
   Area from 0 to $z = 0.63$ is 0.2357.
   Subtract from half the area under the curve (0.5).
   Therefore area to the right of $z = 0.63$ is 0.2643.

4. Find the area to the left of $z = 1.65$ (see Figure 2.9).
   Area from 0 to $z = 1.65$ is 0.4505.
   Add to area for half the curve (0.5).
   Area to the left of $z = 1.65$ is 0.9505.

These calculations have been related mainly to positive z values and to the positive half of the curve, but the same principles are involved in dealing with negative z values. In all cases a rough diagram is a great help in ensuring that z values are correctly assigned to their position on the curve.

close to 0.5, its limiting value. Technically, therefore, the normal curve can be treated as a closed curve, since there is negligible addition to total frequency at values of z higher than 5 or lower than -5.

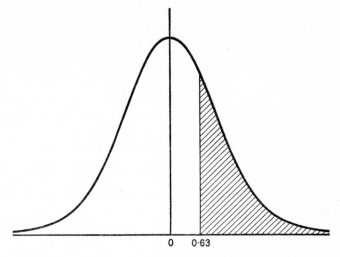

Figure 2.8

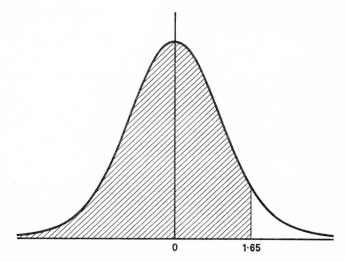

Figure 2.9

2.4 Use of the standard normal distribution

To sum up, the standard normal distribution is a symmetrical frequency distribution. The values of the variable z are measured along the horizontal axis, and the area under the curve, as in any histogram or frequency polygon, represents total frequency. Since total frequency is taken as 1 (akin to total probability), the area under the standard normal curve is treated as unity. The area between any two z values measured along the horizontal axis therefore represents the proportionate frequency with which items occur between those two values.

Suppose that the height of each boy in a group of 50 eleven-year-old boys is measured, and that the mean height is $143 \cdot 25$ cm and the standard deviation is $2 \cdot 75$ cm. This type of distribution would approximate closely to a normal distribution so that the properties of the standard normal distribution can be used to answer the following types of question:

1.   What proportion, or number, of boys are between 140 cm and 145 cm in height?
2.   What proportion, or number, of boys are (a) $137 \cdot 5$ cm tall or less, (b) $147 \cdot 5$ cm tall or taller?
3.   How tall are the tallest 10% of the boys? And the shortest 20%?

For questions 1 and 2 the first step is to turn the actual heights into z values.

Thus $x_1 = 140$ cm corresponds to

$$z_1 = \frac{x_1 - \mu}{\sigma}$$

i.e. $z_1 = \dfrac{140 - 143 \cdot 25}{2 \cdot 75} = -1 \cdot 18.$

For $x_2 = 145$ cm,

$$z_2 = \frac{145 - 143 \cdot 25}{2 \cdot 75} = 0 \cdot 64.$$

Figure 2.10 shows the z values, and the area between them represents the proportion of boys between 140 cm and 145 cm tall.

Area from 0 to z_1 $(-1 \cdot 18) = 0 \cdot 3810.$
Area from 0 to z_2 $(0 \cdot 64) = 0 \cdot 2389.$
Area between z_1 and $z_2 = 0 \cdot 6199.$

This shows that approximately $0 \cdot 62$ or 62 per cent of all the eleven-year-old boys in the group would be expected* to be be-

* In this context 'expected' refers to a 'mathematical expectation'. It does not imply that in any random sample of 50 eleven-year-old boys with the given mean and standard deviation, 62% or 31 boys must *necessarily* be between 140 cm and 145 cm tall. The mathematical expectation is that if a very large number of such groups of 50 boys were obtained, then *on average* the number of boys

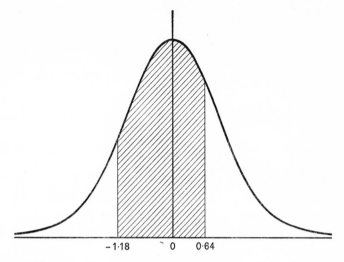

-1·18   0   0·64

Figure 2.10

tween 140 cm and 145 cm tall. Alternatively, out of 50 boys, 31 would be expected to fall within this range.

For boys 137·5 cm tall or less,

$$z_3 = \frac{137 \cdot 5 - 143 \cdot 25}{2 \cdot 75} = -2 \cdot 09.$$

The normal curve area table gives areas from the mean (0) to the given z value. Here what is required is the area in the tail of the curve (see Figure 2.11).

Area under half curve $= 0 \cdot 5000$.
Area in half curve to z_3 $(-2 \cdot 09) = 0 \cdot 4817$.
Area in tail from z_3 $= 0 \cdot 0183$.

This shows that approximately $0 \cdot 02$ or 2% of boys would be expected to be 137·5 cm in height or less, and the expected number in a group of 50 boys would therefore be 1.

For boys 147·5 cm in height or more,

$$z_4 = \frac{147 \cdot 5 - 143 \cdot 25}{2 \cdot 75} = 1 \cdot 55.$$

between 140 cm and 145 cm tall would be 62% or 31 boys. A simple analogy is with coin-tossing. If a balanced coin is tossed 50 times, then the 'mathematical expectation' is that 25 results would be heads uppermost and 25 tails uppermost. However, the result for a single experiment of tossing a coin 50 times would not necessarily be to get exactly 25 heads and 25 tails. See also section 3.1.

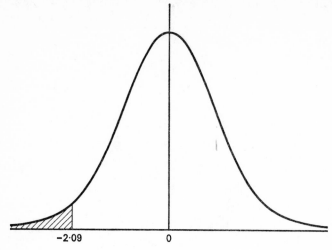

Figure 2.11

Again the area representing frequency is given in the tail of the curve, to the right of $z_4 = 1 \cdot 55$ (see Figure 2.12).

Area in half curve $= 0 \cdot 5000$.
Area from 0 to z_4 $(1 \cdot 55) = 0 \cdot 4394$.

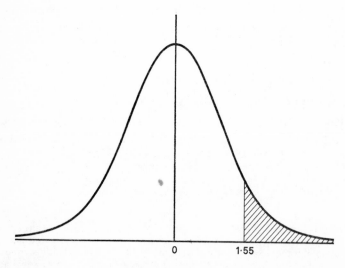

Figure 2.12

Area in tail from $z_4 = 0 \cdot 0606$.

Approximately $0 \cdot 06$ or 6% of the boys would therefore be expected to be at least $147 \cdot 5$ cm tall, that is, three boys in a group of fifty.

The last question requires the z value to be found which corresponds to a particular area. For the tallest 10 per cent, z_5 corresponds to an area of $0 \cdot 1$ in the right hand tail, that is, an area of $0 \cdot 4$ from 0 to z_5. Similarly, for the shortest 20 per cent, z_6 corresponds to an area of $0 \cdot 2$ in the left hand tail, an area of $0 \cdot 3$ from 0 to z_6 (see Figure $2 \cdot 13$). Looking up in the body of the normal curve area table, for areas $0 \cdot 4$ and $0 \cdot 3$, $z_5 = 1 \cdot 28$ and $z_6 = -0 \cdot 84$ (since it is in the negative half of the curve).

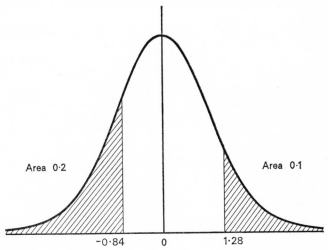

Area 0·2

Area 0·1

−0·84   0   1·28

Figure 2.13

These z values can be converted back into the original units since

$$x_i = \mu + (z_i \times \sigma)$$

Therefore $x_5 = 143 \cdot 25$ cm $+ (1 \cdot 28 \times 2 \cdot 75)$ cm $= 146 \cdot 77$ cm

and $\quad x_6 = 143 \cdot 25$ cm $- (0 \cdot 84 \times 2 \cdot 75)$ cm $= 140 \cdot 94$ cm.

This means that 10 per cent of the boys (or 5 in number) would be expected to be $146 \cdot 77$ cm or more in height, and that 20 per cent (or 10 in number) would be expected to be $140 \cdot 94$ cm or less.

Many sets of data form frequency distributions which

approximate to a normal distribution. Given the mean and standard deviation of the data, any original value can be converted into 'standard units' and related to the properties of the standard normal distribution, through the normal curve area table. Alternatively, standard units relating to specific normal curve areas can be converted into values in terms of the actual data. The properties of the standard normal distribution can in this way be used to examine the properties of a set of data.

2.5   The standard normal distribution and the binomial distribution

As mentioned in section 2.2, the normal distribution can be used in place of a binomial distribution, where the normal distribution is given the same mean and standard deviation as the binomial distribution. The practical use of this is that the calculations involving the normal distribution are very much simpler than those involving the binomial distribution, where all frequencies have to be worked out separately. Consider as an example, the tossing of ten balanced coins in a series of 100 trials. The proportion of trials in which theoretically seven or more heads would be tossed could be obtained by calculating the separate binomial frequencies for 7, 8, 9 and 10 heads. These would be as follows:

$$7 \text{ heads } 120 \times (\tfrac{1}{2})^7 \times (\tfrac{1}{2})^3 = 120 \times (\tfrac{1}{2})^{10}$$
$$8 \text{ heads } 45 \times (\tfrac{1}{2})^8 \times (\tfrac{1}{2})^2 = 45 \times (\tfrac{1}{2})^{10}$$
$$9 \text{ heads } 10 \times (\tfrac{1}{2})^9 \times (\tfrac{1}{2}) = 10 \times (\tfrac{1}{2})^{10}$$
$$10 \text{ heads } 1 \times (\tfrac{1}{2})^{10} \phantom{\times (\tfrac{1}{2})^2} = 1 \times (\tfrac{1}{2})^{10}$$

The total of these frequencies is $176 \times (\tfrac{1}{2})^{10}$, or $0 \cdot 1719$. In a hundred trials, therefore, 17 results would be expected in which 7 or more heads appeared. Equally, in one thousand trials, 172 such results would be expected.

The same result can be obtained using the normal distribution. The mean and standard deviations of a binomial distribution* are given by

$$\mu = np$$

$$\sigma = \sqrt{\{np(1 - p)\}}$$

In this example therefore, $n = 10$ (the number of coins

---

* If the *proportion* of heads were being investigated, as opposed to the actual *number* of heads, the mean of the appropriate binomial distribution would be p, and the standard deviation $\sqrt{\{p(1 - p)/n\}}$

tossed), and $p = \frac{1}{2}$ (the probability of tossing a head). So that $\mu = 5$, the expected mean number of heads when ten coins are tossed, and $\sigma = \sqrt{(10 \times \frac{1}{2} \times \frac{1}{2})}$ or $1 \cdot 581$.

The binomial distribution is now going to be treated as approximating to a normal distribution with mean of 5 and standard deviation of $1 \cdot 581$. Before the standard units (z) value is calculated which corresponds to 7 heads in the binomial distribution, allowance must be made for the discrete nature of the binomial distribution. No value falls between 6 and 7 heads, between 7 and 8 heads, etc. Thus, in conformity with the usual treatment for relating discrete data to a continuous form, the value 6 heads is treated as covering the range $5 \cdot 5$ to $6 \cdot 5$, 7 heads as covering $6 \cdot 5$ to $7 \cdot 5$, 8 heads as covering $7 \cdot 5$ to $8 \cdot 5$, etc. The z value required therefore is that corresponding to $6 \cdot 5$ heads, not 7 heads. Figure 2.14 makes this clearer.

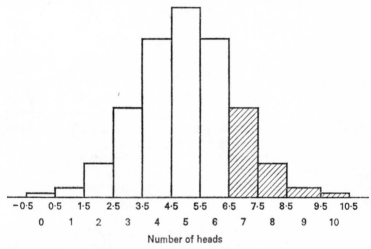

Figure 2.14

For $6 \cdot 5$ heads, $z = (x - \mu)/\sigma$,

that is, $z = \dfrac{6 \cdot 5 - 5}{1 \cdot 581} = 0 \cdot 95$ (see Figure 2.15)

Area in half curve $= 0 \cdot 5000$

Area to $z = 0 \cdot 95$ is $0 \cdot 3289$

Area to right of $z = 0 \cdot 95$ (i.e. in tail) is $0 \cdot 1711$.

This result is very close indeed to the result obtained from using the binomial frequencies, which is $0 \cdot 1719$. The con-

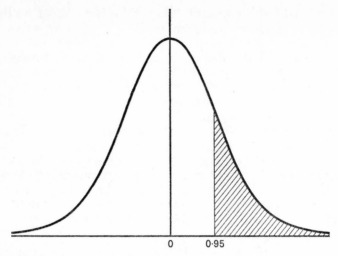

Figure 2.15

clusions would be the same, that in a series of 100 trials in which ten coins were tossed, 7 heads or more would be expected to appear 17 times.

The approximation is closer the larger n is, i.e. the greater the number of trials. Clearly, as n increases, the discrete binomial distribution approximates more and more closely to a continuous distribution—a normal distribution (see Figure 2.16). The approximation continues to be good even where $p \neq \frac{1}{2}$, that is, the probabilities of success and failure are not equal, as in rolling a dice for sixes. The only circumstances where the normal distribution is not a reasonable approximation is when p is either very small or very large (see section 2.6).

It is very much simpler to use the normal distribution than to calculate binomial frequencies, especially as the great majority of instances when these values are required involve values of n much higher than 10, and far more than four binomial frequencies, as in the example quoted above.

It is, therefore, possible to make use of the normal curve properties in circumstances in which, strictly speaking, the theoretical distribution is binomial.

2.6 The Poisson distribution

The Poisson distribution is a discrete distribution similar in type to the binomial distribution. Where p, the probability of success (or failure) is very small, the distribution of the results of

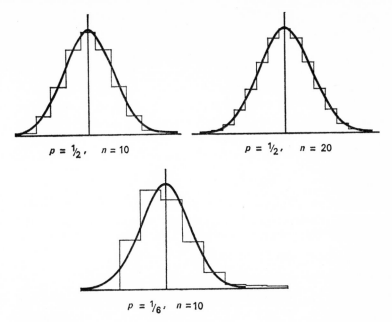

$p = \frac{1}{2},\quad n = 10$ $\qquad\qquad$ $p = \frac{1}{2},\quad n = 20$

$p = \frac{1}{6},\quad n = 10$

Figure 2.16

a series of trials would approximate to the Poisson distribution, not to the binomial distribution.

The most useful form of the Poisson distribution is given by:

$$P(x) = \frac{\mu^x}{x!} exp\,(-\mu)$$

where $P(x)$ is the proportionate frequency, or probability of x successes in n trials. $\mu = np$, p being the probability of success in an individual trial, and n the number of trials, as for the binomial distribution. $Exp\,(-\mu)$ is the same as $\dfrac{1}{exp\,(\mu)}$ so that, although it is not normally written this way, the expression could be rearranged as:

$$P(x) = \frac{\mu^x}{x!\,exp\,(\mu)}$$

Again, as for the binomial distribution, the frequency for each specific number of successes has to be calculated separately. This makes the Poisson distribution distinctly tedious to use, but calculations can be helped by using a table of logar-

D

ithms of factorials.* Fortunately, the circumstances in which it should be used are very restricted, and its application is therefore of limited importance. The approximation of the Poisson distribution to the binomial distribution improves as np ($=\mu$) increases, and this approximation is close for values of $np = 5$ or more. Thus, provided np is at least 5 (i.e. where, although p the probability is small, n the number of trials is large), the binomial distribution is appropriate, and in its turn, the normal distribution can be used in place of the binomial.

Exercises

(Section 2.3)

1.   A set of measurements has a mean of $56 \cdot 5$ cm and a standard deviation of $1 \cdot 7$ cm. Convert each of the following measurements into standard units:

(*a*) $60 \cdot 75$ cm   (*b*) $51 \cdot 74$ cm   (*c*) $75 \cdot 37$ cm

(*d*) $56 \cdot 16$ cm   (*e*) $63 \cdot 81$ cm   (*f*) $43 \cdot 75$ cm.

For each part of questions 2 and 3, a rough diagram is an essential aid:

2.   Given the following z values, find the normal curve area which lies

(*a*) to the right of $z = 2 \cdot 2$

(*b*) to the left of $z = 0 \cdot 85$

(*c*) to the right of $z = -1 \cdot 76$

(*d*) to the left of $z = -1 \cdot 24$

(*e*) between $z = -1 \cdot 4$ and $z = 0 \cdot 23$

(*f*) between $z = 0 \cdot 74$ and $z = 1 \cdot 33$

(*g*) between $z = -1 \cdot 55$ and $z = -0 \cdot 37$.

3.   Given the following normal curve areas, derive z, if

(*a*) the normal curve area between 0 and z is $0 \cdot 3770$

(*b*) the normal curve area to the left of z is $0 \cdot 1112$

(*c*) the normal curve area to the right of z is $0 \cdot 4207$

(*d*) the normal curve area to the left of z is $0 \cdot 6664$

(*e*) the normal curve area to the right of z is $0 \cdot 9945$

(*f*) the normal curve area between $-z$ and $+z$ is $0 \cdot 8030$.

* A table of logarithms of factorials is included in D. V. Lindley and J. C. P. Miller, *Cambridge Elementary Statistical Tables* (Cambridge University Press, 1953).

(Sections 2.4 and 2.5)

4. A group of 1,000 doctors has a mean age of $47 \cdot 35$ years and a standard deviation of $10 \cdot 92$ years. On the assumption that their age distribution approximates to a normal distribution

(a) how many doctors are under 40 years of age?

(b) how many doctors are under 60 years of age?

(c) what age excludes the youngest 15 per cent?

(d) what age excludes the oldest 10 per cent?

5. The mean number of heads and the standard deviation of the binomial distribution which is obtained by tossing 20 coins are given by $\mu = 10$ and $\sigma = 2 \cdot 24$. What is the probability, when making a single toss, of obtaining

(a) *exactly* 10 heads?

(b) 15 or more heads?

(c) fewer than 4 heads?

(d) between 8 and 11 heads?

6. If ten dice are rolled for sixes, the mean number of sixes and the standard deviation of the appropriate binomial distribution are given by $\mu = 1 \cdot 67$ and $\sigma = 1 \cdot 18$. What is the probability, in a single roll, of obtaining

(a) no sixes at all?

(b) one six?

(c) 2, 3 or 4 sixes?

(d) 5 or more sixes?

Further reading

Useful discussions of one or more of the binomial, normal and Poisson distributions may be found in:

A. Bradford Hill, *Principles of Medical Statistics* (The Lancet, 8th edition 1966), chapter 7.
H. T. Hayslett, *Statistics Made Simple* (W. H. Allen, 1968), chapters 5 and 6.
M. J. Moroney, *Facts from Figures* (Penguin Books, 1951), chapters 7, 8 and 9.
W. J. Reichmann, *Use and Abuse of Statistics* (Penguin Books, 1964), chapter 15.

3

The theoretical basis of sampling

3.1 Sample and population

The purpose of taking a sample and investigating its characteristics, say in terms of a sample mean, is to find out something about the population itself. It is necessary, therefore, to start by examining the relationship between sample means and the mean of the parent population where the latter is already known.

Suppose that a sample is drawn from a population consisting of all the students in a college and that for each sample unit a variable, height, is measured. The sample units drawn from a population must reflect the way in which the variable occurs in the population. The variable, height, measured for a random sample of primary-school children, would be quite different from the student heights. In this context there is a fundamental difference between the two populations, namely, that primary school children are shorter than students. This difference between the populations must appear also in the samples. Similarly, the sample means, as central representative values, will in their turn depend upon the characteristics of the populations from which the samples are drawn. Differences between sample means will reflect differences between populations.

If the student population from which the sample is drawn consists entirely of tall students, then the sample itself will consist entirely of tall students, with a correspondingly high sample mean. But if the population contains students varying from the very tall to the very short, the sample, too, can consist of members with widely varying heights. The sample represents the population, so that the variation or spread in the parent population must reappear in the sample. The greater the variation within the population, the greater the variation amongst the sample measurements themselves.

Differences between a population mean and the mean of a sample drawn from it are likely to be small where the population itself shows little variation. On the other hand, where the variation within the population is great, larger differences are likely between the sample mean and the population mean.

The sample mean as a central measure is also equivalent to a centre of gravity with values above and below the mean offset against each other. If the sample contains a disproportionate number of extreme values, these will in turn affect the sample mean and increase the difference between the sample mean and the population mean. The larger the sample, however, the smaller the risk of this distortion by extreme values. Suppose, for example, that the population of students consisted equally of tall and short students (however defined). A sample of only three students might quite easily consist only of short students, but if the sample were increased in size to ten or twenty students, then clearly it would be very much less likely that the larger samples would be made up entirely of short students. This is equivalent to saying that if three coins are tossed it is quite possible all might come down heads. The probability of this happening is $(\frac{1}{2})^3 = 0 \cdot 125$. It is very unlikely that if ten coins were tossed all would be heads since the probability here is $1/1024$ or $0 \cdot 001$; and even more unlikely if twenty coins were tossed, with a probability of less than one in a million that twenty heads would occur.

The same sort of arguments apply to sample and population proportions. If a coin is tossed, then (on assumption that it is a balanced coin) in half the tosses it should come down heads and in the other half tails. That is, the expected proportion of heads is $0 \cdot 5$. If in 100 tosses of a coin, the proportion of heads was not exactly $0 \cdot 5$, but fairly close, say between $0 \cdot 45$ and $0 \cdot 55$ (representing 45 to 55 heads), this would not be looked upon as an unreasonable result. But if the proportion of heads was only, say, $0 \cdot 3$ (producing 30 heads in 100 tosses), the result would be looked upon as extreme. The coin might not be balanced or the tosses might not be unbiased.

Suppose the experiment were to be repeated a number of times, then a frequency distribution of the sample results, the proportion of heads in 100 tosses of a coin, could be constructed. The pattern of the distribution would be such that most of the sample proportions would be clustered around (above and below) the population proportion. There would be some more extreme high and low proportions, but these would occur relatively infrequently.

The points so far mentioned are concerned with a common-sense appreciation of the relationship between a sample statistic (whether mean or proportion) and the population parameter to which it is related. In the case of the relationship between the sample mean and its population mean, this section can be summarized by saying that

(i) the greater the variation (standard deviation) in the population, the greater the variation within a sample drawn from it, and potentially the larger the difference between sample mean and population mean;

(ii) the larger the sample size, given the variation in the population, the smaller will be the difference between the population mean and the sample mean.

In the case of proportions, the situation is slightly different, since variations between sample proportion and population proportion depend in part on the magnitude of the population proportion itself. However, the sample size bears the same relationship to variation as for sample mean and population mean;

(iii) the larger the sample size, given the population proportion, the smaller will be the difference between the population proportion and the sample proportion.

3.2 Experimental sampling distributions: means

Suppose that information is available about the heights of all members of a population of 900 students. The heights are grouped to form the distribution shown in Table 3.1.

Table 3.1 *Heights of 900 students*

Height (cm)	Number of students
160 and under 164	31
164 and under 168	122
168 and under 172	154
172 and under 176	175
176 and under 180	197
180 and under 184	136
184 and under 188	65
188 and under 192	16
192 and under 196	4
All heights	900

The mean height of this population is $174 \cdot 64$ cm and the standard deviation is $6 \cdot 86$ cm.

Suppose that, by random methods, the units in the population are divided into 90 groups each of 10 units. Although this cannot strictly be looked upon as random sampling, since effectively sampling is without replacement and this places an absolute limit on the total number of samples that can be drawn, the situation provides a sufficiently close approximation to random sampling for the purpose of the present argument. The sample means are then calculated for each group of 10 and these means can themselves be formed into a frequency distribution as in Table 3.2.

Table 3.2   *Mean heights of samples of 10 students*

Mean height (cm)	Number of samples
169 and under 170	3
170 and under 171	2
171 and under 172	4
172 and under 173	9
173 and under 174	18
174 and under 175	16
175 and under 176	13
176 and under 177	13
177 and under 178	8
178 and under 179	2
179 and under 180	1
180 and under 181	1
All heights	90

This is known as an experimental sampling distribution, since it is obtained by the experimental drawing of samples. In this particular case, the mean of the sampling distribution must equal the population mean, since each sample unit appears in one sample only. The mean of the sampling distribution is therefore 174·64 cm. The standard deviation of the sampling distribution is 2·14 cm, which is much smaller than the standard deviation of the population itself. Clearly, the variation between sample means is very much less than the variation within the population.

If the groups are now taken together in pairs, so that the sample size is doubled from 10 to 20, a new experimental sampling distribution is obtained, which is shown in Table 3.3.

Table 3.3   *Mean heights of samples of 20 students*

Mean height (cm)	Number of samples
171 and under 172	1
172 and under 173	4
173 and under 174	11
174 and under 175	14
175 and under 176	6
176 and under 177	6
177 and under 178	3
All heights	45

Again, the mean of this distribution is equal to the population mean, but its standard deviation is 1·44 cm. This is smaller than the standard deviation of the previous sampling distribution, when sample size was 10. The actual frequency distribution covers a narrower range of values, and the spread, or variation, within this experimental sampling distribution is clearly less than in the previous distribution when sample size was 10.

Finally, if the samples are regrouped in threes to provide 30 samples each of 30 students, yet another experimental sampling distribution is formed (see Table 3.4), with the same mean as the population and the two previous experimental sampling distributions, but with a yet smaller standard deviation, 1·34 cm. Once more, the variation in the data has been reduced with increased sample size.

The general conclusion is, therefore, that as sample size has increased, variation, as measured by the standard deviation of the sampling distribution, has fallen. This point is examined further in section 3.4 below.

3.3   Experimental sampling distributions: proportions

The student population consisted of both men and women, 480 men and 420 women. The proportion of men in the student population was therefore 480/900 = 0·533. In a sample of 100 students, the expected number of men would be 53; in a sample of 10 students, the expected number of men would be 5 (in both cases to the nearest whole number). Instead of measuring

Table 3.4 *Mean heights of samples of 30 students*

Mean height (cm)	Number of samples
172 and under 173	3
173 and under 174	8
174 and under 175	7
175 and under 176	8
176 and under 177	2
177 and under 178	2
All heights	30

the heights of a sample of students, and calculating the mean height for the sample, the number of men in each sample group could have been recorded. As mentioned in section 3.1, samples would be unlikely to consist entirely of men, or entirely of women. On the other hand, it would be surprising if a sample showed exactly the same proportion of men as in the parent population. What would be expected would be a sample proportion which was close to the population proportion, but not necessarily identical with it.

Table 3.5 *Number of men in 90 samples of 10*

Number of men in sample	Number of samples	Proportion of men in sample
0	0	0·0
1	2	0·1
2	2	0·2
3	8	0·3
4	16	0·4
5	21	0·5
6	18	0·6
7	14	0·7
8	6	0·8
9	3	0·9
10	0	1·0
All samples	90	

Suppose that, as in section 3.2, the student population is divided by random methods into groups of 10 students and the number of men in each group recorded. The results are set out in Table 3.5.

The mean of this distribution of sample proportions is $0 \cdot 533$, the same as the population proportion, and the standard deviation is $0 \cdot 171$.

If the samples are randomly paired to make 45 samples of 20 students (see Table 3.6), the new experimental sampling distribution has a mean proportion of $0 \cdot 533$, but this time the standard deviation is $0 \cdot 107$.

Table 3.6   *Number of men in 45 samples of 20 students*

Number of men in sample	Number of samples	Proportion of men in sample
5	0	$0 \cdot 25$
6	1	$0 \cdot 30$
7	3	$0 \cdot 35$
8	2	$0 \cdot 40$
9	6	$0 \cdot 45$
10	11	$0 \cdot 50$
11	8	$0 \cdot 55$
12	4	$0 \cdot 60$
13	5	$0 \cdot 65$
14	3	$0 \cdot 70$
15	2	$0 \cdot 75$
16	0	$0 \cdot 80$
All samples	45	

The variation within the distribution of sample proportions has been reduced as the sample size has increased.

Finally, if the data is regrouped to form 30 samples of 30 students, the mean proportion remains as before at $0 \cdot 533$, and the standard deviation is yet further reduced to $0 \cdot 103$. The results follow the pattern discussed in section 3.1 above. The standard deviation of the sampling distribution is progressively reduced as sample size increases.

3.4   Theoretical sampling distributions

What is now necessary is to establish the formal relationship between the population and the sampling distribution.

The argument in section 3.2 above is somewhat limited, because the sampling distribution is based on a very small number of random samples. If, however, an infinitely large number of random samples of a given size were drawn from the student population, and the mean heights calculated, the effect would be to obtain an experimental sampling distribution which would approximate very closely indeed to a normal distribution. The normal distribution would have a mean equal to the population mean μ and a standard deviation equal to $\sigma/\sqrt{n}$, where σ is the population standard deviation and n the sample size. The larger the number of random samples taken, the closer would be this approximation. This relationship between the standard deviation of the theoretical sampling distribution and the standard deviation of the population from which the samples are drawn can be derived mathematically; it is not dependent solely on an empirical relationship.

On this basis, the standard deviation of the theoretical sampling distribution when $n = 10$ is $2 \cdot 16$ cm $(= 6 \cdot 86 \, \text{cm}/\sqrt{10})$ compared with the experimental value of $2 \cdot 14$ cm. For the second experimental sampling distribution when $n = 20$, the calculated value of the standard deviation is $1 \cdot 44$ cm, compared with the theoretical value $1 \cdot 54$ cm. And for the third distribution, the calculated value of the standard deviation is $1 \cdot 34$ cm compared with the theoretical value of $1 \cdot 26$ cm. The discrepancies are due to the relatively small numbers of samples in the experimental distributions.

The close approximation between an experimental sampling distribution and the normal distribution holds either when the sample is drawn from a population which is itself normally distributed, as in the example of the variable height in the student population, or when the sample is large, where 'large' means thirty or more items. The present discussion is in these terms. The special case of the sample of less than thirty drawn from a population which does not approximate to a normal distribution is dealt with in section 4.6.

For a theoretical sampling distribution of proportions, the mean proportion is equal to the population proportion θ,* and the standard deviation is $\sqrt{\{\theta(1-\theta)/n\}}$, where n is the sample size. This mean and standard deviation are the mean proportion and standard deviation of a binomial distribution with probability of success θ and sample size n. The theoretical sampling

* The Greek letter θ (pronounced 'theta') is here used for the population proportion. Some texts use π, but as this Greek letter already has a well-established usage as the ratio of the circumference of a circle to its diameter, a different Greek letter is used so that no confusion can arise.

distribution approximates very closely indeed to a normal distribution with this mean and standard deviation (see section 2.5).

Thus for a sample size of 10 students where $\theta = 0.533$, the mean proportion for the sampling distribution will also be 0.533, and the standard deviation of the theoretical sampling distribution, $\sqrt{(0.533 \times 0.467/10)}$, or 0.158. This compares with the calculated value from the experimental sampling distribution of 0.171. With samples of 20 students the standard deviation is theoretically 0.111, compared with the experimental value of 0.107; and finally, for samples of 30 students, the theoretical value of the standard deviation is 0.091, as compared with the experimental value of 0.103.

Although the theoretical and calculated experimental values of the standard deviations are not identical, they are certainly close. The discrepancies are the result of the relatively small numbers of samples making up the experimental distributions. If the number of samples making up the experimental distributions were to be increased further, the calculated values would tend to move more closely towards the theoretical values.

3.5 Sample statistic and population parameter

If the population characteristics are known, then, since a normal distribution is completely determined by its mean and standard deviation, the properties of the theoretical sampling distribution are also known. The normal distribution was discussed in chapter 2, and there the point was made that any normal distribution can be turned into a common form by giving it a common mean and standard deviation, conventionally a mean of 0 and a standard deviation of 1. Values are measured not as $x_1, x_2, x_3, \ldots, x_n$, but as $z_1, z_2, z_3, \ldots, z_n$, where $z_i = (x_i - \mu)/\sigma$. For the sampling distribution, with standard deviation $\sigma/\sqrt{n}$,

$$z_i = \frac{x_i - \mu}{\sigma/\sqrt{n}}.$$

Thus z is the difference of the original value from the mean, expressed in units of standard deviation (see section 2.3).

The case of sample proportions is the same as for sample means. The z value is here the difference between the single sample proportion p, and the mean proportion of the sampling distribution, which in turn is equal to the population proportion θ. That is,

$$z_i = \frac{p_i - \theta}{\sqrt{\{\theta(1 - \theta)/n\}}}$$

the difference between the sample proportion and the population proportion, divided by the standard deviation of the sampling distribution.

As in any histogram or frequency polygon, the area under the normal curve represents frequency, and a proportion of the area represents proportionate frequency. Here, the frequency distribution is the theoretical sampling distribution of sample means.

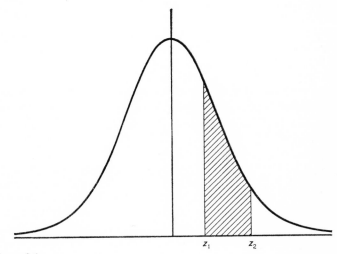

Figure 3.1

If the area between z_1 and z_2 in Figure 3.1 contains a fraction, say $0 \cdot 2$, of the total area under the normal curve, then between z_1 and z_2 lies $0 \cdot 2$ of total frequency. Further, the normal distribution can be looked upon as a probability distribution. If $0 \cdot 2$ of total frequency lies between z_1 and z_2, then there is a $0 \cdot 2$ probability that any single item out of total frequency will have a value lying between z_1 and z_2. In the context of sampling, this is an important concept, namely that proportionate frequency represents probability.

Returning to the theoretical sampling distribution, the properties of the normal distribution can be used to examine the relationship between sample means and the population mean, the latter itself being the mean of the sampling distribution. The standard deviation here referred to is the standard deviation of the sampling distribution, not the population.

Figure 3.2 shows the approximate percentages (proportions) of the total area under the normal curve which falls between certain values. These values are taken, for convenience,* sym-

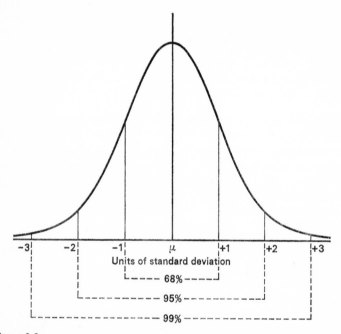

Figure 3.2

metrically round the mean as ±1 standard deviation, ±2 standard deviations and ±3 standard deviations.

Thus, making use of what is effectively a probability distribution, it can be inferred that:

68% of sample means will lie within the population mean ±1 standard deviation,

95% will lie within the population mean ±2 standard deviations,

99% will lie within the population mean ±3 standard deviations.

---

* In later chapters, a more precise relationship between normal curve areas and their related *z* values is used.

As a corollary, there is a probability of 0·68 (68%) that a single sample mean will lie within the population mean ± 1 standard deviation, a probability of 0·95 (95%) that it will lie within the population mean ± 2 standard deviations, and a probability of 0·99 (99%) that it will lie within the population mean ± 3 standard deviations.

In terms of proportions, 68% of sample proportions will lie within the range of the population proportion ± 1 standard deviation, 95% within the population proportion ± 2 standard deviations, and so on. And for the single sample proportion, there is a probability of 0·68 that it will lie within the population proportion ± 1 standard deviation, 0·95 probability that it will lie within the population proportion ± 2 standard deviations, etc.

Thus with varying degrees of probability and within varying limits, the single sample mean or proportion can be related to the corresponding parameter of the population from which it is drawn. This depends on knowing already the appropriate characteristics of the population. For the sample mean this would be the population mean and standard deviation. For the sample proportion, this would be the population proportion. But in real life, population parameters are usually unknown, and the problem is one of inferring something about the unknown population characteristics in terms of the sample results. This is considered in the following chapter.

Further reading

The relationship between sample and population, and sampling distributions, are discussed in:

A. Bradford Hill, *Principles of Medical Statistics* (The Lancet, 8th ed. 1966) chapters 9 and 10.

S. M. Dornbusch and C. F. Schmid, *A Primer of Social Statistics* (McGraw-Hill, 1955) chapters 9 and 10.

J. E. Freund and F. J. Williams, *Modern Business Statistics* (Pitman, 2nd ed. 1970) chapter 8.

M. J. Moroney, *Facts from Figures* (Penguin Books, 1951) chapter 10.

A number of interesting illustrations of the relationship between the distribution of a variable in the population and the corresponding sampling distribution of the mean are contained in: F. E. Croxton, D. J. Cowden and S. Klein *Applied General Statistics* (Pitman, 3rd ed. 1968) chapter 24.

Estimation

4.1 Introduction

Chapter 3 was concerned with the relationship between the sampling distribution and the population from which the samples were drawn. This was extended to considering a single sample statistic (the mean, and then the proportion) and its relationship in probability terms to its population parameter. It was implicit throughout chapter 3 that the population and its relevant characteristics were known. The questions which now arise and which are dealt with in this and the following chapter, are concerned with the situation where nothing is known about the population, and the only information available pertains to a single sample.

This indeed is a much more common actual situation than that discussed in chapter 3. In social and other sample surveys, in clinical trials, in scientific and psychological experiments, the data obtained can only be a sample drawn from a much larger and unknown population. The conclusions reached about the sample or samples are ultimately intended to apply to the population as a whole. The next step, therefore, is to consider how far sample results can be looked upon as being representative of the population.

4.2 The single sample: the mean

Consider a sample of 50 students, drawn from the student population used in chapter 3, for whom the calculated mean height is 174·94 cm, and the standard deviation 6·42 cm. If the population mean is unknown, the sample mean itself would constitute the sole available information relating to the population mean. The sample mean is therefore used as the 'best estimate' of the population mean, since no other and no better information is available. But it is clear from chapter 3 that although the value of a sample mean is likely to be fairly close to the population mean, the two will not necessarily be identical (and indeed are not, since the population mean is known to be 174·64 cm). What is needed, as well as this single 'best estimate', is some means of expressing the possible variation which may arise between this 'best estimate' and the true population mean.

The argument of chapter 3 can be turned round. Given that the sample mean is 174·94 cm, the mean height of the population is likely to be fairly close to this sample value, not greatly dissimilar from it, but not necessarily identical with it. A sample with a given mean could be drawn from a number of different populations, that is, populations whose means would lie clustered round the given sample mean. The argument may seem tortuous, but these population means can be looked upon as being normally distributed round the sample mean, having a distribution mean equal to the sample mean, and with the same standard deviation as the sampling distribution ($\sigma/\sqrt{n}$; see section 3.4). In chapter 3 it was argued that the probability of obtaining a given mean could be derived from the properties of the normal distribution as applied to the sampling distribution. Given the sample mean, referred to as $\bar{x}$, the probabilities can be derived for ranges of possible population means, provided something is known about the population standard deviation.

The variation in the sample (and hence its standard deviation also) depends on the variation in the population. If nothing is known about the population, then this information is not available. In such circumstances the variation in the sample can be looked upon as a measure of the variation in the population since it derives from it. There are, however, slight difficulties in using s, the standard deviation of the sample, as a 'best estimate' of σ, since the sample standard deviation is biased and underestimates the population standard deviation. The larger the sample size, the closer is the approximation between the sample and population standard deviations. As n gets progressively larger, so the sample standard deviation approaches the population standard deviation. For practical purposes, the sample standard deviation is sufficiently close to the population standard deviation, when the sample size is at least 30, for this discrepancy to be ignored. This section, therefore, applies when n, the sample size, is large, here defined as 30 or more. The case of smaller samples is dealt with in section 4.6.

The standard deviation s for the mean height of the sample of 50 students is known, and is 6·42 cm. This can be used as a 'best estimate' of the population standard deviation σ. This, therefore, provides the basis of an estimate of the standard deviation of the sampling distribution $\sigma/\sqrt{n}$. s is used in place of

E

σ, and $s/\sqrt{n} = 6\cdot42\ \text{cm}/\sqrt{50} = 0\cdot91\ \text{cm}$. The standard deviation of the dispersion of possible population means round the sample mean is therefore $0\cdot91$ cm for the sample of 50 students with mean height $174\cdot94$ cm.

Making use of the properties of the normal distribution, it can be asserted that given the mean and standard deviation of the sample, of the possible population means,

approximately 68% lie between the sample mean $\pm\ s/\sqrt{n}$
　　　　　95% lie between the sample mean $\pm\ 1\cdot96\ s/\sqrt{n}$
　　　　　99% lie between the sample mean $\pm\ 2\cdot58\ s/\sqrt{n}$
　　　　　99·9% lie between the sample mean $\pm\ 3\cdot3\ s/\sqrt{n}$.

This is shown in figure $4\cdot1$.

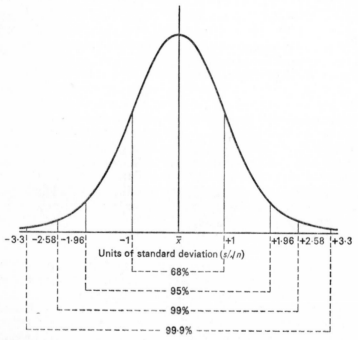

Figure 4.1

The population mean, therefore, is estimated as the sample mean, but the ranges of value expressed above give an indication of the possible variation from this 'best estimate'. Since approximately 68% of possible population means fall within the range of the sample mean $\pm\ s/\sqrt{n}$, and 95% within the sample mean $\pm 1.96 \, s/\sqrt{n}$ and so on, there is a 0.68 probability that the population from which the sample is actually drawn will have a mean within the range $\bar{x} \pm s/\sqrt{n}$, a 0.95 probability that it will have a mean within the range $\bar{x} \pm 1.96 \, s/\sqrt{n}$ and so on.

For the students, therefore, the population mean is estimated as 174.94 cm. In addition, the possible variation between the actual population mean and its 'best estimate' of 174.94 cm can be expressed by saying that *of all populations from which the sample could be drawn*

> approximately 68% have means within 174.94 cm ± 0.91 cm
> or 174.03 cm to 175.85 cm
>
> 95% have means within 174.94 cm ± 1.78 cm
> or 173.16 cm to 176.72 cm
>
> 99% have means within 174.94 cm ± 2.35 cm
> or 172.59 cm to 177.29 cm
>
> 99.9% have means within 174.94 cm ± 3.00 cm or 171.94 cm to 177.94 cm.

This can also be expressed in probability terms by saying that there is, for example, a 0.95 probability that the population mean height lies between 173.16 cm and 176.72 cm, and a probability of 0.99 that it lies between 172.59 cm and 177.29 cm.

As previously mentioned, these conclusions only apply when n, the sample size, is large, here defined as 30 or more. The case of smaller samples is dealt with in section 4.6.

4.3 The standard error

The standard deviation of the sampling distribution is a measure of the probable difference between the sample mean and the population mean (see section 4.4). Alternatively, it is a measure of the probable error in using the sample mean as an estimate of the population mean (see section 4.5 below). For this reason, the standard deviation of a sampling distribution is referred to as the 'standard error', or as the 'standard error of the estimate', or as the 'standard error of the mean' (or proportion, or other measure). This emphasizes the nature and purpose of the standard deviation of the sampling distribution, that is, its function in measuring the error of the sample statistic as an estimate of the population parameter. In the remainder of this book, the standard deviation of a theoretical sampling distribution will be referred to as the standard error.

4.4 Probability and confidence limits

The frequency distribution of population means referred to in section 4.2 can be looked upon as a probability distribution. There is a probability of 0·95 that the population mean lies within the range of sample mean ± 1·96 standard errors. Conversely, there is a 0·05 probability that it does not, a 0·05 probability that the population mean lies outside (below or above) these limits of value. Similarly, in saying that there is a 0·99 probability that the population mean lies within the sample mean ± 2·58 standard errors, there is a 0·01 probability that this is not so—that the population mean lies outside these limits.

Thus the information about the sample of 50 students could be expressed in these terms by stating that there is a 0·95 probability that the population mean lies between 173·16 cm and 176·72 cm, or a 0·99 probability that it lies between 172·59 cm and 177·29 cm.

In this way the extent of the range of values is related to the probability of being correct in inferring that this range contains the population mean. The converse is that the range does not contain the population mean. In statistical methods involving samples, there is no certainty, only probability. It is worth remembering what probability means in this context. If a very large number of exercises were carried out, each based on a single sample result, and ranges of value for sample mean ± 1·96 standard errors calculated in each case, then in 0·95 instances, or 95 out of 100, the estimate of the population mean would lie within these limits. In the remaining 0·05 or 5 per cent of cases, the population mean would lie below the lower limit, or above the higher limit, that is outside the quoted range for the population mean. For 0·99 probability, the range is greater (sample mean ± 2·58 standard errors), but the probability of being outside these limits is less, at 0·01, or 1 in 100 cases. This increase in the probability of being correct is at the price of extending the range of values within which it is claimed the population mean will fall. Increased reliability is at the price of a loss of precision in the result.

These ranges of value are commonly referred to as 'confidence limits'. For example, in estimating the student population mean height as 174·94 cm from the sample of 50 students, the method used allows a 95% confidence that the actual population mean lies between 173·16 cm and 176·72 cm; and a 99% confidence that it lies between 172·59 cm and 177·29 cm.

Confidence in the sense in which it is used here is confidence in the statistical methods used and in the results thereby obtained. In the long run, 95% of the population means *will* lie within the range of the sample mean $\pm$ 1·96 standard errors, and 99% within the sample mean $\pm$ 2·58 standard errors.

Confidence limits, or alternatively, levels of probability, can be drawn at any point that is desired. Confidence limits of any magnitude can be obtained. The coefficient of the standard error is the standard units value, or z value, for the proportion of area under half the normal curve which corresponds to the desired confidence limits. For the 99·9% confidence limits (0·001 probability), the z value corresponding to an area under half the normal curve of 0·4995 is 3·3. The 99·9% confidence limits are therefore given by sample mean $\pm$ 3·3 standard errors (see Figure 4.2).

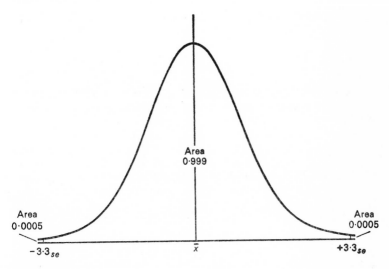

Figure 4.2

The use of specific limits (chiefly 95%, 99% and 99·9%) has developed as a matter of practical convenience. These limits have been found to 'work'; that is, they produce meaningful results. For data in which a high degree of precision is unlikely, and most social and economic data fall into this category, 95% confidence limits work in practice. A greater degree of certainty may sometimes be required, in which case, say, 99% confidence limits might be used. For data of a more scientific nature, in which expected variations would be slight anyhow, and where a higher degree of certainty is usually desirable, higher confidence limits, 99% or above, would be used.

4.5 Probable error

Confidence limits at a given probability denote a range of values within which the population mean is likely to fall. An alternative way of expressing the same result is to state the probable error in using the sample result as the estimate for the population mean. The 0·95 probable error is therefore $\pm$ 1·96 $\times$ standard error, and the 0·99 probable error is $\pm$ 2·58 $\times$ standard error.

In terms of the sample of 50 students, the mean height of the population is estimated to be 174·94 cm. The 0·95 probable error is $\pm$ 1·78 cm (= 1·96 $\times$ 0·91 cm) and the 0·99 probable error is $\pm$ 2·35 cm (= 2·58 $\times$ 0·91 cm).

4.6 The mean of small samples

The statistic

$$z = \frac{\bar{x} - \mu}{\sigma/\sqrt{n}}$$

where $\bar{x}$ is the individual sample mean, n the sample size and μ and σ the population mean and standard deviation respectively, is normally distributed. But the statistic

$$t = \frac{\bar{x} - \mu}{s/\sqrt{n}}$$

where s, the sample standard deviation, is being used as a 'best estimate' of σ, the population standard deviation, is not distributed normally. As was mentioned in section 4.2 above, s, and hence $s/\sqrt{n}$, itself varies with the sample size n. The appropriate theoretical distribution for the t statistic is known as the Student-t distribution. The name, Student, is taken from the pseudonym under which W. S. Gossett published his investigations of this distribution in 1908.

The Student-t distribution is also a symmetrical distribution, slightly flatter and more extended than the normal curve. Effectively, there is a separate t distribution for every value of n. Tables showing the value of t relate sample size, in terms of what are known as 'degrees of freedom',* and given probability

---

* N.B. The important concept, the number of 'degrees of freedom', is the number of ways in which a set of data can vary, given the restrictions imposed by the nature of the calculation to be performed. In calculating s, deviations round the mean are taken. But their sum $\sum_{i=1}^{n} (x_i - \bar{x}) = 0$. Thus if $(n - 1)$ deviations are known, the last is automatically determined. The number of degrees of freedom inherent in the calculation of t is therefore $(n - 1)$.

levels. The flatter and more extended shape of the Student-*t* distribution ensures that a higher proportion of its area is in the tails of the distribution as compared with the normal distribution. This means that at given probability levels, values of *t* are higher than the corresponding values of *z*. Also, the smaller *n*, the sample size, the fewer the degrees of freedom, and the larger the *t* values.

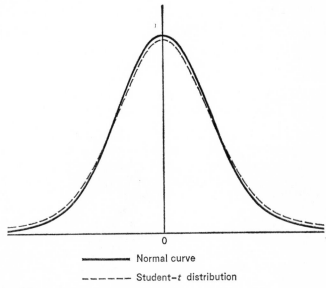

0

———— Normal curve

— — — — — Student-*t* distribution

Figure 4.3

The significance of size 30 as the dividing line between small and large samples is that at this sample size, the Student-*t* distribution approximates very closely to the normal distribution, and variations due to sample size can be ignored (see Figure 4.3).

For sample size below 30, confidence limits are wider, and probable errors larger, than for samples of more than 30 items. As the sample size falls, the confidence limits become progressively wider, and the probable errors progressively larger.

The net effect may be that for a very small sample, the confidence limits are so wide, or the probable error of the estimate of the mean so large that any conclusion is of very little real use.

One of the samples of 10 students referred to in chapter 3 was selected by random means and was found to have a mean height of 176·10 cm and a standard deviation of 3·88 cm. For 9 degrees of freedom ($n - 1$, where $n = 10$) and probability 0·05 (that is, 0·025 in each 'tail' of the appropriate t distribution),

$$t_{0.05} = 2·262.*$$

The 95% confidence limits for the mean height of the student population would be given by:

$$\bar{x} \pm ts/\sqrt{n} \text{ or } 176·10 \pm 2·262 \times 3·88/\sqrt{10} \text{ cm.}$$

This gives limits of 176·10 ± 2·78 cm, 173·32 cm to 178·88 cm, which are clearly wider than the corresponding limits, 173·16 cm to 176·72 cm, obtained from the sample of 50 students.

4.7 The single sample: the proportion

The same arguments as in section 4·2 hold in relating the sample proportion and the population proportion. A sample with a given proportion could have been drawn from any of a number of populations with different population proportions. The probability of the sample having been drawn from a population with a specified proportion will be determined by the theoretical sampling distribution. As explained in section 2.5, this sampling distribution is strictly a binomial distribution, but it can for practical purposes be taken to be a normal distribution with the same mean and standard deviation as the binomial distribution. The mean is the sample proportion p, and the standard deviation, which is here the standard error, is $\sqrt{\{p(1 - p)/n\}}$.†

Using the normal distribution properties, of the possible population proportions from which a sample with given proportion, p, could be drawn:

approximately 68% will lie between $p \pm 1$ standard error

95% will lie between $p \pm 1·96$ standard errors

99% will lie between $p \pm 2·58$ standard errors.

* The subscript to $t_{0.05}$ refers to the area in the two tails of the distribution excluded by the value $t = \pm 2·262$. It is the value of t, for 9 degrees of freedom, corresponding to $z = 1·96$, where 0·05 is excluded in the two tails of the distribution taken together. See Appendix, Table II. Again, as for the normal distribution, it corresponds to the situation where 0·95 of total area lies between the two t values.

† Since nothing is known about the population, the sample proportion p is used as the 'best estimate' of θ, the population proportion. The standard error, therefore, is expressed in terms of p.

In the sample of 50 students drawn from the student population, 29 were male. This gives a sample proportion of 0·58 males. The mean and standard deviation (standard error) of the appropriate sampling distribution are:

mean (proportion) = 0·58
standard error   = $\sqrt{0\cdot58 \times 0\cdot42/50} = \sqrt{0\cdot004872}$
                   = 0·0698 or 0·070.

In terms of the possible population proportions of male students:

approx. 68% will lie between 0·58 ± 0·070 (0·510–0·650)
  95% will lie between 0·58 ± 0·137 (0·443–0·717)
  99% will lie between 0·58 ± 0·181 (0·399–0·761).

Again, there is a probability of 0·95 that the population proportion of male students would lie between 0·443 and 0·717, or a probability of 0·99 that it would lie between 0·399 and 0·761.

Alternatively, the sample proportion, 0·58, is used as the 'best estimate' of the population proportion, since this is the only information available. The degree of accuracy is then expressed through the probable error. The population proportion is estimated as 0·58; the 0·95 probable error on this estimate is 0·137, and the 0·99 probable error is 0·181.*

4.8 Error and sample size

In practice, the magnitude of the error in a population estimate is often important. Results may be required in terms of a specific range of values (confidence limits), and a less precise population estimate outside the desired range could invalidate

---

* Technically, the standard error of the proportion depends on the unknown population proportion θ. That is, the limits of value of θ, the population proportion, are obtained from the roots of the equation $p = \theta \pm k\sqrt{\{\theta(1 - \theta)/n\}}$ where k is the critical value or standard units value for the probability in question, e.g. 1·96 for 0·95 probability.

The roots of this equation are given by:

$$\theta = \frac{np + \frac{1}{2}k^2 \pm k \sqrt{\{np(1 - p) + \frac{1}{4}k^2\}}}{n + k^2}$$

In the case of the sample of 50 students, for probability 0·95 and $k = 1\cdot96$, the limits of value are $\theta = 0\cdot442$ and 0·706. These compare with the slightly wider 95% confidence limits above, 0·443–0·717. The approximation obtained by using p as the best estimate of θ is reasonably close, and improves as n increases in size. It is clearly a much simpler calculation than that involved in calculating the limits of p in terms of θ.

the whole research or investigation. As has already been seen in chapter 3, standard errors are reduced as sample size increases. If a more precise result is required, then the sample size must be increased. Equally, the larger the sample, the greater the cost of the project. It is not economic, therefore, to conduct an investigation that produces results of a higher degree of accuracy than required.

From section 4.5, probable error $= k \times$ standard error, where k is the standard units value corresponding to the acceptable probability (e.g. $1 \cdot 96$ for $p = 0 \cdot 05$).

If the acceptable probable error is E, for the given probability, then for projects where estimates of population mean are required,

$$E = ks/\sqrt{n}, \text{ so that } n = k^2 s^2/E^2.$$

Therefore, if something is known about the standard deviation, either of the sample (from previous studies or a pilot survey) or the population (from, for example, census figures), a proposed sample size n, for acceptable error E, can be calculated.

Similarly, where estimates are to be made of the population proportion,

$$E = k\sqrt{\{p(1 - p)/n\}}, \text{ so that } n = k^2 p(1 - p)/E^2.$$

Again, if some estimate of either the sample or population proportion is available, n can be calculated for a given value of E. In this case, an estimate of p is not essential. The expression $p(1 - p)$ has a maximum value when $p = 0 \cdot 5$, and $p(1 - p) = 0 \cdot 25$. Substituting in the expression above, $n = k^2/4E^2$.

A sample of size n therefore will produce an error not exceeding E, irrespective of the value of p.

4.9 Sampling from small populations

All the preceding discussions of error have implied that it is the size of the sample and not the size of the population which affects the magnitude of the standard error. For by far the majority of practical purposes, this is correct. The calculations of standard errors have assumed infinitely large populations. Some populations from which samples are drawn clearly have no limit; for example, samples derived from tossing pennies or rolling dice. More often, however, populations are finite, although their size may not be known. Where a population is finite, and its magnitude known, the calculation of the standard error can take this into account. In practice this is only done where the sample represents a relatively large fraction of the population. The correction to the standard error in such circumstances is:

$$\mathrm{se}_{\bar{x}} = \frac{s}{\sqrt{n}} \times \sqrt{\left\{\frac{N-n}{N-1}\right\}} \text{ and } \mathrm{se}_p = \sqrt{\left\{\frac{p(1-p)}{n}\right\}} \times \sqrt{\left\{\frac{N-n}{N-1}\right\}}$$

where N = population size, and n = sample size.

Suppose, for example, that $N = 1,000$, and $n = 50$, so that the sample consists of 5% of the population units. The correcting factor $\sqrt{\left\{\frac{N-n}{N-1}\right\}}$ is 0·978. The effect of taking the population size into account is to reduce the standard error by 2·2%. If the sample size was 100, i.e. 10% of a population of 1,000, the correction for the small population would be 0·949, and the standard error would have been reduced by 5·1%.

It is clear that very little reduction in the standard error is obtained by this correction unless a large fraction of the population is being sampled. Any decision whether or not to take it into account is inevitably arbitrary. As a rough guide, the operation of the small population correction will not make a real difference to the estimate of the standard error where the proportion of the population sampled is less than 5%.

4.10 Summary and notes on calculations

The foregoing sections of this chapter set out the basis for estimating population parameters from sample results. This can be summarized as follows:

1. The sample statistic is used as a 'best estimate' of the population parameter.
2. The reliability of the sample statistic as the 'best estimate' of the population parameter can only be expressed in probability terms, either in relation to confidence limits or to probable error. It is essential that some qualification should be made to the population estimate, in view of the differences which can and do arise between sample statistics and population parameters.

The following examples demonstrate the different ways in which reliability may be expressed.

1. A sample of 500 randomly selected children aged 9 and under 10 years were given a particular reading test. The mean score was 68·9, and the standard deviation 3·6. Estimate (*a*) the 95% confidence limits, and (*b*) the 99% confidence limits for

the population mean score for all children aged 9 and under 10 years.

$$\text{se}_{\bar{x}} = s/\sqrt{n} = 3 \cdot 6/\sqrt{500} = 0 \cdot 16$$

The 95% confidence limits are given by

$$\mu = \bar{x} \pm 1 \cdot 96 \times \text{se}_{\bar{x}}$$
$$= 68 \cdot 9 \pm 1 \cdot 96 \times 0 \cdot 16$$
$$= 68 \cdot 9 \pm 0 \cdot 32$$
$$= 68 \cdot 58 \text{ to } 69 \cdot 22.$$

The 99% confidence limits are given by

$$\mu = 68 \cdot 9 \pm 2 \cdot 58 \times 0 \cdot 16$$
$$= 68 \cdot 9 \pm 0 \cdot 41$$
$$= 68 \cdot 49 \text{ to } 69 \cdot 31.$$

There is, therefore, a $0 \cdot 95$ probability that the population mean score of all children 9 years and under 10 at the given time lies between $68 \cdot 58$ and $69 \cdot 22$; and a $0 \cdot 99$ probability that the population mean lies between $68 \cdot 49$ and $69 \cdot 31$.

2.   In a random sample of 100 dwellings owned by the local housing authority in a particular town, the mean number of habitable rooms per dwelling was $3 \cdot 25$, and the standard deviation was $1 \cdot 2$ rooms. From this data (*a*) obtain the 95% confidence limits on the estimate of the mean number of habitable rooms in all dwellings owned by the local authority in the town, and (*b*) express the $0 \cdot 99$ probable error in estimating the population mean number of habitable rooms per dwelling as $3 \cdot 25$.

The standard error of the mean is

$$\text{se}_{\bar{x}} = s/\sqrt{n} = 1 \cdot 2/\sqrt{100} = 0 \cdot 12$$

The 95% confidence limits are given by

$$\mu = \bar{x} \pm 1 \cdot 96 \, \text{se}_{\bar{x}}$$
$$= 3 \cdot 25 \pm 1 \cdot 96 \times 0 \cdot 12$$
$$= 3 \cdot 25 \pm 0 \cdot 235$$
$$= 3 \cdot 015 \text{ to } 3 \cdot 485.$$

The mean number of habitable rooms per dwelling is estimated as $3 \cdot 25$, with a $0 \cdot 95$ probability that the population mean lies between $3 \cdot 015$ and $3 \cdot 485$ habitable rooms per dwelling. The $0 \cdot 99$ probable error is given by $E = \pm 2 \cdot 58 \times \text{se}_{\bar{x}}$.

That is, $E = \pm\ 2\cdot58 \times 0\cdot12$
$\qquad = \pm\ 0\cdot31.$

In estimating the population mean number of habitable rooms as $3\cdot25$, there is a $0\cdot99$ probable error of $\pm\ 0\cdot31$ habitable rooms.

3. In a survey of 225 businessmen conducted through a manufacturers' association, 45% $(0\cdot45)$ thought that trade prospects would improve during the next six months, 32% $(0\cdot32)$ thought they would get worse, and 23% $(0\cdot23)$ thought there would be no change either way. On the assumption that these views were expressed by a random sample of business men, construct (*a*) 95% confidence limits, and (*b*) 99% confidence limits for the proportions thinking that trade prospects would improve and trade prospects would get worse respectively.

For the group who thought trade prospects would improve, $p = 0\cdot45$, and the standard error of p, $\sqrt{\{p(1 - p)/n\}}$, is given by

$\mathrm{se}_p = \sqrt{(0\cdot45 \times 0\cdot55/225)} = \sqrt{0\cdot0011}$
$\qquad = 0\cdot0332.$

The 95% confidence limits are given by

$\theta = p \pm 1\cdot96 \times \mathrm{se}_p$
$\quad = 0\cdot45 \pm 1\cdot96 \times 0\cdot0332$
$\quad = 0\cdot45 \pm 0\cdot065$
$\quad = 0\cdot385$ to $0\cdot515.$

The population proportion is estimated as $0\cdot45$ or 45%; it represents the proportion of all businessmen who think trade prospects will improve, and the 95% confidence limits are $0\cdot385$ to $0\cdot515$ or $38\cdot5\%$ to $51\cdot5\%$.

Similarly, the 99% confidence limits are given by

$\theta = p \pm 2\cdot58 \times \mathrm{se}_p$
$\quad = 0\cdot45 \pm 0\cdot085$
$\quad = 0\cdot365$ to $0\cdot535.$

The 99% confidence limits for the population mean indicate that there is a $0\cdot99$ probability that of all businessmen, the proportion who think trade prospects will improve lies between $0\cdot365$ or $36\cdot5\%$, and $0\cdot535$ or $53\cdot5\%$.

For the businessmen who thought trade prospects were

getting worse, $p = 0.32$ and $se_p = \sqrt{(0.32 \times 0.68/225)} = 0.0311$.

The 95% confidence limits are therefore given by $0.32 \pm 1.96 \times 0.0311$ or 0.259 to 0.381. The 99% confidence limits are given by $0.32 \pm 2.58 \times 0.0311$, or 0.240 to 0.400. The proportion of all businessmen who think trade prospects are getting worse is estimated at 0.32 or 32%, with 95% confidence that the true figure lies within the range 0.259 to 0.381 (25.9% to 38.1%) and 99% confidence that the true figure lies between 0.246 and 0.400 (24.0% to 40.0%).

4. In a market research survey of a random sample of 700 housewives, 460 stated that they preferred to use tea-bags rather than loose tea for tea making; the remainder preferred to use loose tea. Calculate (a) the 0.95 probable error in estimating the proportion of all housewives who prefer to use tea-bags as 0.66, and (b) the 0.99 probable error.

The 0.95 probable error is given by

$E = \pm 1.96 \times se_p$

$E = \pm 1.96 \times \sqrt{\{p(1-p)/n\}}$ where $p = 460/700 = 0.66$

$E = \pm 1.96 \times \sqrt{(0.66 \times 0.34/700)} = \pm 1.96 \times 0.0179$

$\quad = \pm 0.0358.$

The 0.99 probable error is given by $2.58 \times se_p$

$E = \pm 2.58 \times 0.0179 = \pm 0.0463.$

In estimating the proportion of all housewives (i.e. the population proportion) who prefer using tea-bags as 0.66 or 66%, there is a 0.95 probable error of ± 0.0358 or 3.58%, and a 0.99 probable error of ± 0.0463 or 4.63%.

Note on working standard errors of proportions in percentage terms

All references to standard errors of proportions have been in terms of decimal fractions of 1. This is mathematically more satisfactory, and is clearly related to the probability arguments where total probability equals 1. However, in working out examples, readers may prefer to work in percentages, since much data is given in percentage form, and because this avoids the arithmetical confusion which sometimes arises in the handling of very small numbers, all less than 1, in calculating the standard error. If working is to be done in percentage data, care must be taken to ensure that *all* data are turned into the percentage form. The formula for the standard error of the proportion then becomes:

$$\mathrm{se}_p = \sqrt{\{p\%(100 - p)\%/n\}}$$

and the result is, of course, itself a percentage.

Exercises
(Section 4.2)

1. A random sample of 549 men employed full-time in construction work, had a mean weekly income of £22·25 with a standard deviation of £6·60. Express the 95% and the 99% confidence limits for mean weekly income of all men employed full-time in construction.

2. The mean age of head of household for a random sample of 3,704 householders was 51·07 years, with standard deviation 16·98 years. What does the data enable you to say about the mean age of all heads of households?

3. (*a*) From the data in Table 4.1, calculate the mean number of workers per household and also the standard deviation.

(*b*) On the assumption that this is a random sample, obtain the 95% confidence limits for the mean number of workers per household.

Table 4.1 *Family Expenditure Survey 1968: number of workers in household*

Number of workers	Number of Households	% of Households
None	1,356	18·9
One	2,724	37·9
Two	2,250	31·3
Three	613	8·5
Four	186	2·6
Five	42	0·6
Six or more	13	0·2
All households	7,184	100·0

(Section 4.5)

4. A random sample of 525 male students had a mean

height of 181·00 cm with standard deviation 3·50 cm. What is the 0·95 probable error in estimating the mean height of all male students as 181·00 cm?

5.   The mean size of a random sample of 7,184 households was 2·95 persons with standard deviation 1·52 persons. What conclusions can be drawn about the error in estimating the mean size of all households as 2·95 persons?

(Section 4.6)

6.   A random sample of households included 24 families with 5 or more children, in which the mother was working. These families had a mean weekly income of £40·10 and a standard deviation of £13·40. Estimate the mean weekly income of all households of working married women with 5 or more children.

7.   The mean weekly earnings of a random sample of 26 full-time women employees in East Anglia was £13·10 with a standard deviation of £2·80. What is the 0·95 probable error in estimating the mean weekly earnings of all full-time women employees in East Anglia as £13·10?

(Section 4.7)

8.   In a pilot survey of 100 students at a particular university, 88 returned a completed questionnaire within one week of receiving it. On the basis that these students were a random sample, express the 95% and the 99% confidence limits for the proportion of all the students at the university who would return the completed questionnaire within one week of receiving it.

9.   Out of a random sample of 260 students sitting a particular degree examination, 221 passed at their first attempt. What is the 0·95 probable error in estimating the proportion of all students passing this examination at their first attempt as 8·5%?

10.   2,179 households, out of a random sample of 7,184 households were renting unfurnished local authority accommodation. What are the 0·95 and 0·99 probable errors in estimating the proportion of all households renting unfurnished local authority accommodation as 0·303 (30·3%)?

11.   25% of a random sample of 457 male manual workers employed full-time in transport were earning £18·00 per week or less in 1968. What can be deduced about the proportion of all full-time male manual workers employed in transport earning £18·00 per week or less?

12.   4,709 households in a random sample of 7,184 households recorded expenditure on alcoholic drink. Express 99·9% confidence limits for the proportion of all households making expenditure on alcoholic drink during the period of the survey.

Further reading

Discussions of the problems of estimation from sample results are contained in:

F. Conway, *Sampling, an Introduction for Social Scientists* (Allen and Unwin, 1967), chapters 9 and 11.
J. E. Freund and F. J. Williams, *Modern Business Statistics* (Pitman, 2nd ed. 1970), chapter 8.
W. J. Reichmann, *Use and Abuse of Statistics* (Penguin Books, 1964), chapter 15.

F

5

Significance tests

5.1 Introduction

The title of this chapter coincides with the title of this book, not because significance tests are its sole concern, but because they are the principal subject matter, to which the rest relates. In a sense, the preceding chapters have all been leading up to this point. Sampling theory has so far been developed to infer something about the whole population from a single sample result. There is, however, another series of problems which can be answered through sampling theory. Could a sample with mean $\bar{x}$ (or proportion p) have come from a particular population with mean μ (or proportion θ)? Could two samples with means $\bar{x}_1$ and $\bar{x}_2$ (or proportions p_1 and p_2) have been drawn from the same population? If there is an observed difference, $(\bar{x}_1 - \bar{x}_2)$ between two sample means, ostensibly from the same population (or similarly between sample proportions, $p_1 - p_2$) could the difference be accounted for by sampling error? If not, can the difference be measured in any way? These questions are related, and can themselves be turned into other forms; forms in which the development of ideas from sampling theory is similar.

To many students, the operation of tests of these kinds, and the understanding of the underlying theory, represents a key point in their statistical methods course. The purpose of this and the following two chapters is to deal fully with these aspects and to develop logical methods of conducting what are known as significance tests.

5.2 Population mean and sample mean

If a population mean is known, the question can arise, whether a particular sample with a different mean could be drawn from that population. In theory, since the normal curve is open-ended, i.e. although coming very close to it, the normal curve never actually meets the x-axis, there is always an infinitely small probability that an infinitely large (or small) sample mean might have been drawn from a particular population. The arguments developed in chapter 4 show that limits can be placed on the sample values that are at all likely to occur and that it is possible, therefore (but only with varying degrees of probability) to determine whether or not a sample mean came from a particular population.

In effect, this is a rearrangement of the arguments about confidence limits. Consider a population with mean μ and a sample with mean $\bar{x}$; there is a $0 \cdot 95$ probability that any sample drawn from that population will have a mean lying between $\mu \pm 1 \cdot 96 \times se_{\bar{x}}$. There is a $0 \cdot 05$ probability that the mean will lie outside these confidence limits. There is a $0 \cdot 99$ probability that a sample will lie between $\mu \pm 2 \cdot 58 \times se_{\bar{x}}$, and a $0 \cdot 01$ probability that it will lie outside the confidence limits. Again, the question can only be considered in probability terms. If $\bar{x}$ lies outside $\mu \pm 1 \cdot 96 \times se_{\bar{x}}$ there is a less than $0 \cdot 05$ probability that the sample with mean $\bar{x}$ comes from the population with mean μ. And if it lies outside $\mu \pm 2 \cdot 58 \times se_{\bar{x}}$ there is a less than $0 \cdot 01$ probability that the sample comes from the given population.

This situation can be formalized through a significance test. Since there is no certainty, but only probability in situations involving statistical inference, a test must be conducted in relation to a particular probability. If the probability chosen is $0 \cdot 05$, then the test will be conducted in such a way that the conclusion that the sample does not come from the population has a probability of at least $0 \cdot 95$ of being right, and a probability not exceeding $0 \cdot 05$ of being wrong. It is to this latter probability that reference is made in significance tests. The probability of being wrong is referred to as the test level. If the probability chosen is $0 \cdot 05$, then the test level is $0 \cdot 05$ probability. Alternatively, this may be referred to as 'testing at $0 \cdot 05$ probability'.

If the sample mean $\bar{x}$ does lie outside $\mu \pm 1 \cdot 96 \times se_{\bar{x}}$, then using $0 \cdot 05$ probability as a test level, the conclusion would be that the sample with mean $\bar{x}$ does not come from the population with mean μ. Here, $0 \cdot 05$ probability reflects the probability of being wrong in the conclusion that the sample does not come from the population.

The implication of a probability level of $0 \cdot 01$ is that the sample will only be rejected as not coming from the population if $\bar{x}$ lies outside the limits $\mu \pm 2 \cdot 58 \times se_{\bar{x}}$. The probability of being wrong in concluding that the sample did not come from the population is only $0 \cdot 01$, or 1 in 100.

What the probability reflects (whatever level is chosen) is the long-run expectation. If a very large number of sample means are examined, and using a probability of $0 \cdot 05$, the conclusion that the sample with mean $\bar{x}$ does not come from the population because $\bar{x}$ does not lie within $\mu \pm 1 \cdot 96 \times se_{\bar{x}}$, would in the long run be correct 95 times out of 100, and wrong 5 times out of 100. The probability chosen is the probability that the conclusion is wrong. Similarly, for probability $0 \cdot 01$, $0 \cdot 001$ etc., the conclusion that the sample does not come from the population because it does not lie within the appropriate confidence limits, will be wrong 1 in 100 times, 1 in 1,000 times, etc.

If, however, the sample mean does not lie outside $\mu \pm 1 \cdot 96 \times se_{\bar{x}}$, then the conclusion cannot be that the sample does not come from the population. This situation is considered in sections 5.4 and 5.5 following.

Clearly, the problem arises—which test level to select for a particular set of data. The answer can only be that any choice is inevitably subjective, and is justified because that test level has been shown to be effective when used for the sort of data in question. For much sociological and economic data, a probability of $0 \cdot 05$ or $0 \cdot 01$ would normally be used. In scientific fields, more stringent testing with smaller probabilities such as $0 \cdot 001$ or less would be common. The practitioner must justify his methods, for there is no simple and absolute answer to this problem.

5.3 The null hypothesis

The question whether or not a sample could have been drawn from a particular population can be examined by means of what is known as a 'null hypothesis'. This is a statement which hypothesizes that there is no real difference between the sample statistic (say mean), and the corresponding population parameter. If, therefore, there is an observed difference, this is assumed to be accounted for by chance, that is, sampling error. The aim then, is to set up a test situation in which the null hypothesis can be examined, and as a result of which the null hypothesis may, or may not, be rejected. If the null hypothesis is rejected, then the observed difference is considered to be 'significant', or 'statistically significant' (to emphasize its dependence on the statistical processes used), and not to be due to chance, that is, to sampling error.

Consider the sample of 50 students, whose mean height was $174 \cdot 94$ cm with standard deviation $6 \cdot 42$ cm, and suppose that the sample was claimed to be drawn from a population whose mean height was 172·50 cm. The observed difference between sample mean and population mean, $(\bar{x} - \mu)$, is (174·94 − 172·50) cm or 2·44 cm. How likely is it that this difference could be due to chance? If the test is to be at 0·05 probability, this is equivalent to asking whether the sample mean falls within or outside of the 95% confidence limits. The easiest way in which to examine this problem is to turn the observed difference into standard units (a z value), by dividing the observed difference by the appropriate standard error (the standard deviation of the sampling distribution). This z value can then be compared with the critical value, determined by the probability selected. Here the probability is 0·05, so the critical value with which to compare the calculated z value is 1·96. If the probability chosen were to be 0·01, then the critical value of z would be 2·58; and for a probability of 0·001, it would be 3·3.

For the 50 students,

$$se_{\bar{x}} = s/\sqrt{n} = 6\cdot42 \text{ cm}/\sqrt{50} = 0\cdot91 \text{ cm}.$$

The appropriate standard units value for the observed difference between the sample mean and the population mean is given by

$$z = \frac{\bar{x} - \mu}{s/\sqrt{n}} = \frac{174\cdot94 - 172\cdot50}{0\cdot91} = \frac{2\cdot44}{0\cdot91} = 2\cdot68.$$

Since 0·05 probability has been chosen, the critical value of z is 1·96. The null hypothesis is rejected, because the z value calculated from the observed difference exceeds 1·96. This means that the null hypothesis is rejected because it has a probability of less than 0·05 of being true. A difference as big as the observed difference (or bigger) has a less than 0·05 probability of occurring by chance, i.e. through sampling error. Indeed, in this example, the calculated z value exceeds 2·58, and the sample value would therefore occur by chance less than 1 in 100 times (0·01 probability).

Figure 5.1 shows the relationship between the population mean and sample mean in terms of the sampling distribution and the critical value of z.

It helps to emphasize that the calculation of the z value from the observed difference establishes the relationship between the two means in terms of the sampling distribution. It provides an immediate indication of the probability with which the observed difference could have occurred. In this respect it is a much more satisfactory approach than actually establishing confidence limits and seeing whether the sample mean falls within these or outside them.

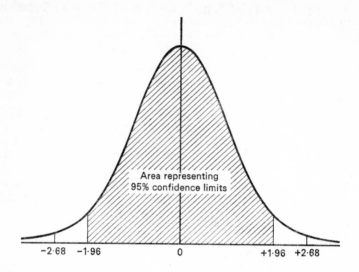

Area representing
95% confidence limits

−2·68 −1·96   0   +1·96 +2·68

Figure 5.1

The test method can be summarized as follows:

1. The null hypothesis is set up, that there is no real difference between $\bar{x}$ and μ, and that any observed difference is due to chance or random sampling error.

2. A suitable probability is chosen, say 0·05. This corresponds to a critical value of z of 1·96.

3. The criterion for rejecting the null hypothesis is established; that is, it will be rejected if $|z| \geqslant z_{0.05} (= 1·96)$,* where the calculated z value is obtained from

$$z = \frac{\text{observed difference}}{\text{appropriate standard error}}$$

4. The z value is calculated; in this example for sample and population mean,†

$$z = \frac{\bar{x} - \mu}{s/\sqrt{n}}$$

5. On the basis of the calculated z value, the null hypothesis is rejected if $|z| \geqslant z_{0.05}$, not rejected if $|z| < z_{0.05}$.

* The symbol $|z|$ refers to the arithmetic value of z, irrespective of sign.

† A negative value of z, obtained from this expression, simply indicates that the sample mean is less than the population mean, and that the sample mean therefore falls in the negative half of the appropriate sampling distribution.

6. If the null hypothesis is rejected, the conclusion is that the sample, mean $\bar{x}$, is not drawn from the population, mean μ. There is a statistically significant difference between the two means.

7. If the null hypothesis is not rejected, then the situation must be considered further. (See sections 5.4 and 5.5 following.)

To revert to the previous example about the fifty students, with mean height 174·94 cm:

1. The probability chosen is 0·05, so that the critical value is $z_{0.05} = 1·96$.

2. The null hypothesis is that there is no real difference between the sample mean height of the group of 50 students, 174·94 cm, and the alleged mean height of the population from which they are drawn, 172·50 cm. Thus the observed difference is due to chance or sampling error.

3. The null hypothesis will be rejected if the calculated $|z| \geqslant z_{0.05}$ that is, if $|z| \geqslant 1·96$.

4. The calculated z value for the difference between the sample and population means is

$$z = \frac{\bar{x} - \mu}{s/\sqrt{n}} = \frac{174·94 - 172·50}{0·91} = \frac{2·44}{0·91} = 2·68.$$

5. $z = 2·68$, therefore $|z| > z_{0.05}$ ($= 1·96$), and the null hypothesis is rejected.

6. The conclusion is, therefore, that at a probability of 0·05 a sample of 50 students with mean height of 174·94 cm could not have been drawn from a population with mean height 172·50 cm.

If the supposed mean of the population from which the students were claimed to have been drawn was, say, 176·00 cm, instead of 172·50 cm, then the mechanics of the test would have been different, only in so far as a different z value would have been calculated for the observed difference.

$$\text{Thus } z = \frac{\bar{x} - \mu}{s/\sqrt{n}} = \frac{174·94 - 176·00}{0·91} = \frac{-1·06}{0·91} = -1·16.$$

Here, the z value is arithmetically smaller than 1·96 (ignoring the minus sign). The first conclusion, as indicated above, is therefore that the null hypothesis cannot be rejected. The question then arises; if the z value is less than the critical value, can the null hypothesis be accepted? A z value of 1·16 corresponds to a probability of 0·2460 that the difference between the sample and population means is due to chance. The probability that the difference is not due to chance is $0 \cdot 7540$. Thus if the null hypothesis is accepted, this has a probability of only $0 \cdot 2460$ of being correct, but a probability of $0 \cdot 7540$ of being wrong (see Figure 5.2).

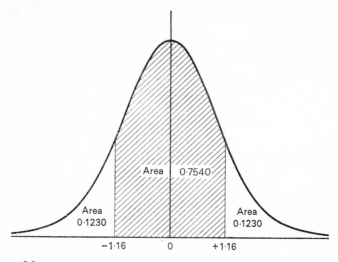

Figure 5.2

As a first word of caution, the alternative to rejecting the null hypothesis is not necessarily to accept it. If it were, then null hypotheses would be being accepted in all instances when the probability that they were correct was in excess of $0 \cdot 05$. For probabilities very little in excess of $0 \cdot 05$, this would clearly be unsatisfactory. This point is considered further in the following sections.

For a small sample, its size n being under 30, the t statistic is calculated and the test conducted similarly against the appropriate critical value of t for $(n - 1)$ degrees of freedom.

5.4   The non-rejection of the null hypothesis

If in a significance test, the calculated z value is very small, the probability that the sample with mean $\bar{x}$ comes from the population with mean μ, is high. For example, if $|z| = 0 \cdot 3$, the probability that a difference as large as $|\bar{x} - \mu|$ (or larger) could by chance occur between the sample and population means is $0 \cdot 7642$ (see Figure 5.3). This is a substantial probability that the sample is in fact drawn from the population.

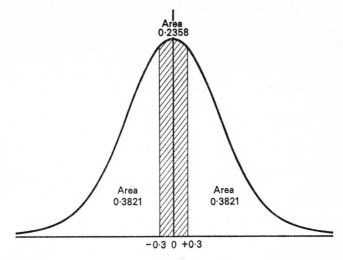

Figure 5.3

The greater the z value, indicating an increasing difference between $\bar{x}$ and μ, the smaller the probability that this difference has occurred by chance and therefore the smaller the probability that the sample is drawn from the population.

Even for higher values, such as $|z| = 1$, the probability of the sample being drawn from the population is quite large. The area in the two tails which represents the probability of the difference $|\bar{x} - \mu|$ being as large as one standard error, or larger, is $0 \cdot 3174$. This is nearly one third of total probability. It is roughly equivalent to the probability of throwing a 5 or 6 with a single throw of a dice.

Thus for small values of z below the critical test value, the null hypothesis can be accepted, at least tentatively. At what values this can be done is again a matter for subjective judgment. As a very rough guide, if $|z| \leqslant 1$, then the null hypothesis might be accepted, since the probability of it being true, approximately one-third, is substantial. This must in any event depend on the nature of the data, and the relative importance of not accepting falsely that the sample comes from the population. The smaller the z value below 1, the greater the firmness that can be expressed in accepting the null hypothesis. A smaller difference is much more likely to be due to sampling error than a larger one. Acceptance can only be on the basis of available information. If this includes a low z value, then there are no grounds for doing other than accepting the null hypothesis.

The real difficulties arise when the z value is relatively high, say over 1, but not as high as the chosen critical test value. This leaves a sort of no-man's land where no real conclusion may be possible. However unsatisfactory this may be, such a non-committal answer may well be the right one. No firm conclusion as to acceptance or rejection of the null hypothesis is not the same as no conclusion at all. It is still positive information.

Such a result indicates the need for additional information, if a firm conclusion is essential. That is, an increase in sample size, or further samples, are required. The effect of increasing the sample size (n) is to reduce the standard error. If the observed difference remains of the same order with increased sample size, the standard error will be reduced, and this will lead to an increase in the calculated z value. The net result would be to increase the likelihood of a positive conclusion. If, however, the observed difference itself changes with increased sample size, the z value may turn out to be sufficiently low to conclude that the samples may well have come from the given population. The first sample would thus prove to have been extreme and unrepresentative.

In setting up a significance test, it is essential to predetermine the criteria which lead to the rejection of the null hypothesis, at a specific critical value for z. It is not so essential to have formulated an alternative hypothesis. In any event, the alternative to rejecting a null hypothesis must not be looked upon *simply* as an acceptance of the null hypothesis.

5.5 Types of error and the alternative hypothesis

This section deals further with the situation which arises when the null hypothesis is not rejected. It is not essential to the remainder of this chapter. Readers who do not wish to pursue further the problem of the alternative hypothesis, or who find this section difficult, can safely pass on to section 5.6.

There are two actual situations which are possible in the relationship between the sample and population means; either the sample comes from the population, or it does not. In the conclusions that can be drawn following a significance test, there are four possible situations, which are set out below:

Null hypothesis: that there is no difference between the sample mean and the population mean; that is, the sample is drawn from the population, and any observed difference is due to sampling error.

The error in concluding that the sample does not come from

Actual situation	Conclusion	
	Sample is from population ↓ ACCEPT NULL HYPOTHESIS	Sample is not from population ↓ REJECT NULL HYPOTHESIS
Sample is from population	RIGHT	WRONG (Type I error— rejecting a true hypothesis
Sample is not from population	WRONG (Type II error— accepting a false hypothesis)	RIGHT

the population when it does (rejecting a true hypothesis) is often referred to as a Type I error. The error in concluding the sample does come from the population when it does not (accepting a false hypothesis) is known as a Type II error. Type I and Type II errors are different and have different implications for the test situation. In many instances, the aim is to avoid, as far as reasonably practicable, a Type I error. But if the alternative to rejecting the null hypothesis is to accept it, in reducing to a minimum level the probability of rejecting a true hypothesis, the risk i.e. probability, is increased that a Type II error will be made. That is, a false hypothesis will be accepted. Although the sample does not come from the population, this is un-recognized.

If a z value is $1 \cdot 96$, or greater, there is a $0 \cdot 05$ probability (or less) that the sample comes from the population. This also means that there is a probability of at least $0 \cdot 95$ that the sample does not come from a population with the given mean. If $0 \cdot 05$ probability is being used for test purposes, and the calculated z value falls just below the critical value of $1 \cdot 96$, say $1 \cdot 75$, then the null hypothesis will not be rejected, because the probability of the sample being drawn from the population is greater than $0 \cdot 05$ ($0 \cdot 08$ for $z = 1 \cdot 75$). If the alternative to rejecting the null hypothesis is to accept it, then what would here be accepted has a probability of only $0 \cdot 08$ of being correct, and a probability of 0.92 of being wrong. The probability of committing a Type II error, therefore, is 0.92.

The basic difficulty is that if the sample mean does not coincide with the population mean, there will be other populations (as determined by their means) from which the sample mean is at least as likely, (or more likely) to have been drawn. And even if the sample mean does coincide with the population mean, there are other populations from which the sample mean $\bar{x}$ is nearly as likely to have been drawn as the hypothesized population with mean μ.

The test situation itself determines the probability of committing a Type I error. But the probability of committing a Type II error arises only when the sample is not drawn from the population and the calculated value of $z, \dfrac{\bar{x} - \mu}{s/\sqrt{n}}$ is less than the critical test value.

Suppose that the sample with mean $\bar{x}$ is drawn from a population whose mean is not μ but μ'; and let the difference between the hypothesized and the actual population means be g standard units. Assume further that the test is at 0.05 probability, for a critical value of $z_{0.05} = 1.96$.

Thus $g = \dfrac{\mu' - \mu}{se_{\bar{x}}}$

Figure 5.4 shows the sampling distribution, for given sample size n, of $\bar{x}$, the sample mean, round the hypothesized population mean μ, and round the actual population mean μ'. In Figure 5.4(a), $g > 1.96$, and in Figure 5.4(b), $g < 1.96$.

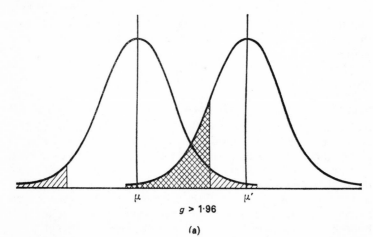

$g > 1.96$

(a)

Figure 5.4(a)

In both (*a*) and (*b*), the areas outside $\mu \pm 1 \cdot 96 \times se_{\bar{x}}$ (hatched) represent the areas for rejection of the null hypothesis at a test probability of $0 \cdot 05$. Within the range $\mu \pm 1 \cdot 96 \times se_{\bar{x}}$ is the area of possible acceptance of the null hypothesis. The possible extent of the Type II error is given by the area of the sampling distribution round the true population mean μ', which falls within the range of possible acceptance of the null hypothesis (cross hatched), that is, within the range $\mu \pm 1 \cdot 96 \times se_{\bar{x}}$.

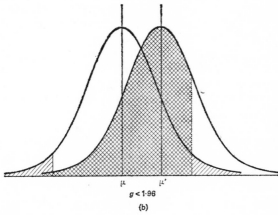

$$\mu \qquad \mu'$$
$$g < 1 \cdot 96$$
(b)

Figure 5.4(*b*)

Thus for $0 \cdot 05$ probability as the test level for the null hypothesis, the probability of making a Type II error is given by the area in the single tail of a normal curve to the left of $z = 1 \cdot 96 - g$.*

Where g is small, the area so described can be large. In any event, where $g < 1 \cdot 96$, the area referred to contains more than half the area under a normal curve, as in Figure 5.4(*b*). This is the sort of situation which arises when the difference between the hypothesized population mean (μ) and the true population mean (μ') is small in relation to the standard error of the mean.

---

* If μ' is less than μ, g will be negative. The probability of making a Type II error would be the area under the normal curve to the right of $z = -1.96 - g$, here g itself being a negative value, so that $-g$ is positive. Thus the situation is exactly equivalent to the one described above when $\mu' > \mu$, and g is a positive value. The significance of 1.96 is that it is the appropriate critical value of z. If the test were to be at $p = 0.01$, then here it would be necessary to use 2.58. The probability of making a Type II error would then be given by the area in the single tail of a normal curve to the left of $z = 2.58 - g$.

In order to estimate the Type II error, therefore, it is necessary either to hypothesize a true population mean, or to stipulate a difference between the hypothesized population mean being used for the significance test, and the true population mean. In one example in section 5.3 above, a sample of 50 students with mean height 174·94 cm and standard deviation 6·42 cm was claimed to be drawn from a population with mean height 176·00 cm. A null hypothesis significance test was conducted, and a calculated z value of $-1·16$ obtained. At 0·05 probability, $|z| < z_{0.05}$ ($= 1·96$) so that the null hypothesis was not rejected. The significance test did not provide any evidence to suggest that the sample was drawn from a population whose mean height was not 176·00 cm. Suppose that the null hypothesis had been positively accepted. The Type II error would have depended on the true population mean height. Suppose that the true population mean height was (i) 174·00 cm, and (ii) 175·60 cm.

For $\mu' = 174·00$ cm,

$$g = \frac{\mu' - \mu}{\mathrm{se}_{\bar{x}}} = \frac{174·00 - 176·00}{0·91} = -2·20.$$

The Type II error is given by the area under the normal curve to the right of

$$z = -1·96 - g = -1·96 + 2·20 = 0·24.$$

This area is 0·405 (see Figure 5.5). Thus the probability of making a Type II error in accepting the null hypothesis that $\mu = 176\cdot00$ cm is 0·405 when the sample is in fact drawn from a population whose true mean height is 174·00 cm.

In the second instance, when $\mu' = 175·60$ cm,

$$g = \frac{\mu' - \mu}{\mathrm{se}_{\bar{x}}} = \frac{175·60 - 176·00}{0·91} = -0·44.$$

Here, the Type II error is given by the area under the normal curve to the right of $z = -1·96 - g = -1·96 + 0·44 = -1·52$.

This area is 0·936 (see Figure 5.6); so that the possible extent of the Type II error in accepting that the sample is drawn from a population with mean height 176·00 cm, when the population mean height is in fact 175·60 cm, is very large at 0·936.

These examples stress the greater magnitude of the possible Type II error when the hypothesized and true population means

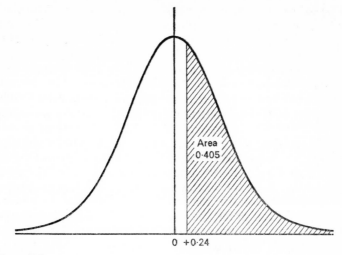

Figure 5.5

are close together. They also indicate that significance tests can be somewhat blunt instruments when it comes to a matter of establishing statistically significant differences in circumstances where relatively small variations are important. If, however, as is quite often the case, it does not really matter whether the student sample is drawn from a population with mean height

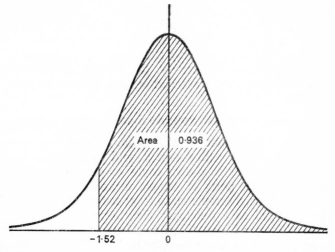

Figure 5.6

176·00 cm *or* 175·60 cm, then the Type II error, though large, may be of no practical importance.

All this argument leads to the conclusion that the alternative to rejecting a null hypothesis is not necessarily to accept it, since to accept it may imply a high risk of a Type II error. The alternative to rejecting the null hypothesis when $|z| < 1·96$ (or other critical test value) requires careful consideration. For small values of z, clearly accepting the null hypothesis is reasonable. At $z = 1·0$, the probability of the sample being drawn from the population is $0·32$—a substantial probability and one which might be considered to justify a positive acceptance of the null hypothesis. But for intermediate values of z, say over $1·0$ and under $1·96$ (or whatever test value is being used), the most realistic alternative may well be to withhold judgment. For z values of this order, the probability of the null hypothesis being true is certainly more than $0·05$ (or other test probability); but the probability of the null hypothesis not being true is substantial.

Clearly, test probability levels, such as $0·05$, $0·01$, $0·001$, are very useful. But they are not a substitute for consideration of the z value actually obtained in a significance test. Not only must careful thought be given to the implications of z values below the critical value for the given probability, but high z values also have their implications. When rejecting a null hypothesis because z exceeds the critical value, say $1·96$, rejection is because the probability of the sample coming from the population does not exceed $0·05$. But the probability of the sample coming from the population may be very much less than $0·05$. At $z = 4·0$, for example, it is only $0·00005$, and at $z = 5·0$, it is only $0·0000006$. Clearly such probabilities are *very* small indeed, and in many situations can be ignored for all practical purposes. Effectively, this sort of situation is as near to certainty as can be reached in the field of sampling.

5.6   Differences between sample means (large samples)

Suppose that a large number of random samples were drawn from two populations, with means μ_1 and μ_2 respectively. The means of the samples so drawn could be calculated, and the differences between pairs of sample means obtained. These differences between a sample mean from each population could then be arranged as a frequency distribution of the statistic $(\bar{x}_1 - \bar{x}_2)$. (Some of the differences between sample means would be positive, others would be negative; the sign of the difference would depend on whether the mean of the sample from the first population was larger than the mean of the sample from the second, or vice versa.)

Such a frequency distribution would form an experimental sampling distribution of the difference between sample means. It would approximate closely to the theoretical sampling distribution of $(\bar{x}_1 - \bar{x}_2)$, which is a normal distribution with a mean of $(\mu_1 - \mu_2)$, and a standard deviation (here the standard error of the difference between sample means) of

$$\sqrt{\left(\frac{\sigma_1^2}{n_1} + \frac{\sigma_2^2}{n_2}\right)}$$

where σ_1 and σ_2 are the standard deviations of the two populations. If, however, these are unknown, then the sample standard deviations s_1 and s_2 can be used as 'best estimates' of the population standard deviations. The standard error then becomes:

$$\mathrm{se}_{\bar{x}_1 - \bar{x}_2} = \sqrt{\left\{\frac{s_1^2}{n_1} + \frac{s_{22}}{n_2}\right\}}.$$

If the samples are drawn from the same, or similar populations, so that $\mu_1 = \mu_2$, then the theoretical sampling distribution would have a mean of 0, $(\mu_1 - \mu_2 = 0)$, and a standard deviation of

$$\sqrt{\left\{\frac{s_1^2}{n_1} + \frac{s_1^2}{n_2}\right\}}.$$

This is the sampling distribution of the difference between means from a single population.

Many situations involve not a single sample mean, which may or may not have come from a particular population, but two sample means which themselves may have come from similar populations. In practical terms, a difference is likely to appear between two sample means, although the samples themselves are drawn from the same or similar populations. The difference between the two sample means may be such that, statistically speaking, the two samples are likely, or are not likely, to have been drawn from the same or similar populations. This leads to the point that the difference between the samples can be investigated in order to throw light on possible differences between the separate populations from which the two samples have been drawn.

If the samples do come from the same population, then theoretically the mean difference between pairs of sample means is itself 0. What must be investigated therefore is the magnitude of the statistic $(\bar{x}_1 - \bar{x}_2)$ to see whether or not its excess over

G

this mean difference (i.e. over 0) could reasonably be attributed to sampling error. The basic approach is exactly the same as for the problem of whether or not a single sample could have come from a particular population. The observed difference between the sample means is turned into standard units and the result compared with an appropriate test value, depending on the probability chosen. If the calculated z value exceeds the test level, then the difference between the sample means is statistically significant; the two samples have not been drawn from the same or similar populations.

To revert to the example of the sample of 50 students with a mean height of 174·94 cm and a standard deviation of 6·42 cm, suppose that a sample of 100 students was drawn from a different college, and that this sample had a mean height of 171·60 cm, and a standard deviation of 6·72 cm. The question to be examined is whether the mean heights of the two student populations could be considered to be the same, that is $\mu_1 = \mu_2$, or whether the sample results suggest that the population mean heights really are different, that is $\mu_1 \neq \mu_2$. The significance test would be carried out as follows:

1.   The probability would be selected for the test, say 0·01.
2.   The null hypothesis is set up that there is no real difference between the mean heights of the population of students at the two colleges; that is $\mu_1 = \mu_2$. Any observed difference between the two sample means, 174·94 cm and 171·60 cm, is therefore due to sampling error.
3.   The null hypothesis will be rejected if $|z| \geqslant z_{0.01}$, that is, if $|z| \geqslant 2 \cdot 58$.
4.   The calculated z value for the difference between sample means is given by

$$z = \frac{\text{observed difference}}{\text{appropriate standard error}}$$

$$= (\bar{x}_1 - \bar{x}_2) \Big/ \sqrt{\left(\frac{s_1{}^2}{n_1} + \frac{s_2{}^2}{n_2}\right)}$$

$$= (174 \cdot 94 - 171 \cdot 60) \Big/ \sqrt{\left(\frac{6 \cdot 42^2}{50} + \frac{6 \cdot 72^2}{100}\right)}$$

$$= \frac{3 \cdot 34}{1 \cdot 13} = 2 \cdot 96$$

5. $z = 2 \cdot 96$, therefore $z > z_{0.01} (= 2 \cdot 58)$ and the null hypothesis is rejected.
6.   The conclusion is that the two samples are not drawn

from the same or similar populations. There is therefore a real difference in the mean height of students at the two colleges. The test has been carried out at a probability of 0·01.

The same considerations about the interpretation of z values apply here, where the difference between sample means is being considered, as have already been discussed in connection with the difference between sample and population means (section 5.2).

5.7 Differences between sample means (small samples)

Where samples are based on small numbers, that is, the sample size is less than 30, the theoretical sampling distribution of the difference between two sample means $\bar{x}_1$ and $\bar{x}_2$, drawn from the sample population, has a mean of 0, and a standard deviation of

$$\sqrt{\left\{\frac{(n_1 - 1)s_1{}^2 + (n_2 - 1)s_2{}^2}{(n_1 - 1) + (n_2 - 1)}\left(\frac{1}{n_1} + \frac{1}{n_2}\right)\right\}}.$$

where n_1 and n_2 are sample sizes, and s_1 and s_2 are sample standard deviations. Thus, $(n_1 - 1)$ and $(n_2 - 1)$ are the degrees of freedom of the two separate samples.

The method of conducting a significance test is exactly the same as in the previous section, but in this case, the appropriate statistic for conducting a significance test is given by:

$$t = (\bar{x}_1 - \bar{x}_2)\Big/\sqrt{\left\{\frac{(n_1 - 1)s_1{}^2 + (n_2 - 1)s_2{}^2}{(n_1 - 1) + (n_2 - 1)}\left(\frac{1}{n_1} + \frac{1}{n_2}\right)\right\}}.$$

This is compared with the Student-t distribution for $[(n_1 - 1) + (n_2 - 1)]$ or $[n_1 + n_2 - 2]$ degrees of freedom.

Suppose two samples, each of 10 students, have mean heights of 176·60 cm and 173·60 cm, and standard deviations of 5·38 cm and 5·62 cm. Could these samples have been drawn from the same population, or is the difference between the two means statistically significant? Conducting the test at 0·05 probability, the null hypothesis is that there is no real difference between the two sample means, but that the observed difference is due to sampling error; the null hypothesis will be rejected if the calculated value of $|t| \geqslant t_{0\cdot05}$ ($= 2\cdot101$ for 18 degrees of freedom).

$$t = (176\cdot60 - 173\cdot60)\Big/\sqrt{\left\{\frac{9\times28\cdot94 + 9\times31\cdot58}{9 + 9}\left(\frac{1}{10} + \frac{1}{10}\right)\right\}}$$

$$= 3\cdot00\Big/\sqrt{\left\{\frac{260\cdot46 + 284\cdot22}{18} \times \frac{2}{10}\right\}}$$

$$= 3 \cdot 00 \Big/ \sqrt{\frac{544 \cdot 68}{90}}$$

$$= \frac{3 \cdot 00}{2 \cdot 46}$$

$$= 1 \cdot 22.$$

The calculated value of t is less than the critical test value ($t_{0 \cdot 05} = 2 \cdot 101$), so that the null hypothesis cannot be rejected. It is, therefore, possible that the two samples came from the same population.

5.8   Population proportion and sample proportion

The considerations here are much the same as for sample means. If a population proportion is known, then the relationship to it of a sample proportion can be examined to decide whether or not a particular sample proportion p is likely to have come from a population with given proportion θ. As has been explained in section 2.5, the theoretical sampling distribution of proportions is strictly binomial, but for practical purposes, the distribution can be taken as approximating very closely to a normal distribution with mean proportion θ, and standard deviation (se_p) of $\sqrt{\{\theta(1 - \theta)/n\}}$ where n is sample size.* As for means, the observed difference between the sample and population proportions can be turned into standard units by means of the expression:

$$z = \frac{\text{observed difference}}{\text{appropriate standard error}} = \frac{p - \theta}{se_p}$$

$$= \frac{p - \theta}{\sqrt{\{\theta(1 - \theta)/n\}}}.$$

Consider a sample of 60 students, of whom 42 are male, who are claimed to be drawn from a student population of whom $0 \cdot 53$ (53%) are male. For the sample, $p = 42/60 = 0 \cdot 70$, and for the population, $\theta = 0 \cdot 53$. Test, at $0 \cdot 01$ probability, the null hypothesis that there is no difference between the proportion of males in the sample of students ($0 \cdot 70$) and the

---

* The standard error of the sample proportion p depends only on θ, the population proportion and on n, the sample size. Where θ is known or hypothesized, the standard error will be in terms of θ. Only where θ is unknown will p be used as a 'best estimate' of θ. In tests concerned with the difference between the sample and population proportions, θ is known and the standard error is expressed in terms of θ.

proportion of males (0·53) in the population from which the students are alleged to be drawn; reject the null hypothesis if $|z| \geqslant z_{0\cdot01}$ (= 2·58).

$$z = \frac{p - \theta}{\sqrt{\{\theta(1 - \theta)/n\}}} = \frac{0\cdot70 - 0\cdot53}{\sqrt{\{0\cdot53 \times 0\cdot47/60\}}} = \frac{0\cdot17}{0\cdot064} = 2\cdot66$$

$z > z_{0\cdot01}$, so that the null hypothesis is rejected. There is a less than 0·01 probability that the difference between the sample proportion and the population proportion could have occurred by chance.

As for tests involving means, the interpretation of the test result depends on the magnitude of z in relation to the critical value. In this particular example, the calculated z value slightly exceeds the test value, $z_{0\cdot01}$. The conclusion is, therefore, that the sample is not drawn from a population with 53% males. If the test had been conducted at 0·05 probability, the null hypothesis would again have been rejected; but if it had been at 0·001 probability, for which the critical value $z_{0\cdot001} = 3\cdot3$, the null hypothesis could not have been rejected, since $z = 2\cdot66$ is less than $z_{0\cdot001}$ (= 3·3). Calculated values of z below the ctitical value would have to be considered carefully, on the basis that the alternative to rejecting the null hypothesis is not necessarily to accept it, unless this is justified by the probabilities implicit in the calculated z value.

5.9 Differences between sample proportions

If there are two populations, with proportions θ_1 and θ_2, and if a series of parallel samples, size n_1 and n_2 respectively, are obtained from the two populations, an experimental sampling distribution can be constructed for the statistic $(p_1 - p_2)$, where p_1 and p_2 are the sample proportions drawn from the two populations. Such an experimental sampling distribution would approximate closely to a theoretical sampling distribution with a mean of $\theta_1 - \theta_2$, and a standard deviation, known as the standard error of the difference between sample proportions, $(se_{p_1 - p_2})$ where

$$se_{p_1 - p_2} = \sqrt{\left\{\frac{\theta_1(1 - \theta_1)}{n_1} + \frac{\theta_2(1 - \theta_2)}{n_2}\right\}}. \qquad (1)$$

This formula is analogous to the formula for the standard error of the difference between sample means $(se_{\bar{x}_1 - \bar{x}_2})$. In both, the standard error is the square root of the sum of the separate squared standard errors. Again, as for sample proportions, the experimental sampling distribution is strictly a binomial distribution, but it corresponds very closely to a normal distribution with the same mean and standard deviation.

In significance test terms, if two sample proportions are claimed to be drawn from similar or the same population, the null hypothesis implies that $\theta_1 = \theta_2 = \theta$. Therefore $\theta_1 - \theta_2 = 0$, so that differences between proportions drawn from the same population tend to cancel out in repeated sampling; again, the same as for sample means. The standard error would become

$$\text{se}_{p_1 - p_2} = \sqrt{\left\{ \theta(1 - \theta)\left(\frac{1}{n_1} + \frac{1}{n_2}\right) \right\}}. \tag{2}$$

In dealing with the difference between two sample proportions, there is no information about the population, or populations, from which the two samples are alleged to be drawn. Thus $\text{se}_{p_1 - p_2}$ can only be expressed in terms of the two sample proportions, p_1 and p_2. In the first version of $\text{se}_{p_1 - p_2}$, p_1 and p_2 can be used as 'best estimates' of θ_1 and θ_2. Thus

$$\text{se}_{p_1 - p_2} = \sqrt{\left\{ \frac{p_1(1 - p_1)}{n_1} + \frac{p_2(1 - p_2)}{n_2} \right\}}. \tag{3}$$

This formula should be used where the two samples are hypothesized as drawn from different populations, and the null hypothesis states that the differences between the two population proportions is of a particular magnitude. This is dealt with in section 6.4.

The second expression above for $\text{se}_{p_1 - p_2}$, which is the more appropriate formula when the two sample proportions are alleged to be drawn from the same population, requires what is known as a 'pooled estimate' for p. In effect, the results of both samples are amalgamated to make a single overall estimate of a combined sample proportion p, which is itself then used as a 'best estimate' of θ, so that

$$p = \frac{n_1 p_1 + n_2 p_2}{n_1 + n_2}.$$

The standard error then becomes

$$\text{se}_{p_1 - p_2} = \sqrt{\left\{ p(1 - p)\left(\frac{1}{n_1} + \frac{1}{n_2}\right) \right\}}. \tag{4}$$

The total *number* of 'successes' from both samples is turned into a proportion of the total number of items for both samples. Where x_1 and x_2 are the number of successes corresponding to the proportions p_1 and p_2, so that

$$p_1 = \frac{x_1}{n_1} \text{ and } p_2 = \frac{x_2}{n_2},$$

$$p = \frac{x_1 + x_2}{n_1 + n_2}.$$

Where the null hypothesis requires that the proportions are drawn from the same population, then the pooled estimate formula for $se_{p_1 - p_2}$ is strictly applicable and mathematically more satisfactory. The standard error calculated from the 'pooled estimate' tends to be slightly larger than that which is calculated from the first formula.

In conducting a significance test, the probability that the difference between the two sampling proportions is due to chance, i.e. sampling error, is obtained as before by calculating a z value:

$$z = \frac{\text{observed difference}}{\text{appropriate standard error}} = \frac{p_1 - p_2}{se_{p_1 - p_2}}.$$

Suppose that $p_1 = 0.55$ and $p_2 = 0.51$ represent the proportions of males in samples of students, size $n_1 = 80$ and $n_2 = 100$, drawn from two different colleges. The question could be asked: are the two populations from which the samples are drawn the same, or similar, in the proportion of male students? If the first sample is drawn from College A and the second from College B, do the samples provide any evidence of a difference between the two colleges in the proportion of male students?

A significance tests would be carried out as follows. First, a suitable probability is chosen, say 0.05. The null hypothesis is set up, that there is no real difference between the two sample proportions of male students at College A and College B; any observed difference is due to sampling error; the colleges are effectively drawn from the same population as far as the proportion of male students is concerned. The criterion for rejecting the null hypothesis is established at $|z| \geqslant z_{0.05} (= 1.96)$, and the value of z calculated from the data as follows:

$$z = (p_1 - p_2) / \sqrt{\left\{ p(1 - p)\left(\frac{1}{n_1} + \frac{1}{n_2}\right) \right\}} \text{ where } p = \frac{n_1 p_1 + n_2 p_2}{n_1 + n_2}$$

$$p = \frac{80 \times 0.55 + 100 \times 0.51}{80 + 100} = \frac{44 + 51}{180} = \frac{95}{180} = 0.528$$

$$z = \frac{0.55 - 0.51}{\sqrt{\left\{ 0.528 \times 0.472\left(\frac{1}{80} + \frac{1}{100}\right) \right\}}} = \frac{0.04}{0.075} = 0.53.$$

$z = 0 \cdot 53$ is less than $z_{0.05}$ so that the null hypothesis cannot be rejected. The calculated z value is in fact very small, so that the null hypothesis could be accepted ($p = 0 \cdot 5962$). Thus the difference between the proportions of males at the two colleges could be due to sampling error, and the claim that there is no real difference between the colleges in this respect can be accepted. The two samples can be looked upon as having been drawn from the same population.

5.10 Two-tailed tests and one-tailed tests

The null hypothesis is that there is no real difference between two values (a sample statistic and a population parameter, or two sample statistics) and that any observed difference is due to chance, that is, to sampling error only.

The alternative hypothesis may be simply that a real non-sampling difference exists between the two values, or it may indicate the direction of the difference. For example, if a manufacturer claims that the contents of a particular pack of biscuits weigh, on average, 225 g, he is claiming a population mean of 225 g. No one is concerned if the mean weight of a number of packets turns out to be 235 g. The manufacturer is simply giving away biscuits. But if the mean weight of a number of packets turns out to be 220 g, doubts would arise as to the correctness of the manufacturer's claim that the mean weight of the contents of all such packets is 225 g.

If what is at issue is simply the magnitude of the difference between two values, then the null hypothesis is in effect rejected if the sample value falls outside the appropriate confidence limits, or if the z value exceeds the critical test value. That is, the null hypothesis is rejected in cases of *extreme* difference in *either direction*, whether the sample statistic exceeds or falls short of the population parameter. For 95% confidence limits, or $0 \cdot 05$ probability, the null hypothesis is rejected for both positive and negative values of z that fall in either tail of the appropriate normal distribution.

This is shown in Figure 5.7, where the shaded area in the two tails together represent $0 \cdot 05$ of total area under the normal curve.

Where, however, direction of difference is important, the area of probability strictly lies in one tail only of the distribution. For the biscuit manufacturer, what matters is whether the z value for the difference between the mean weight of the sample of bisuit packets and the population mean weight he claims falls into the negative tail of the distribution.

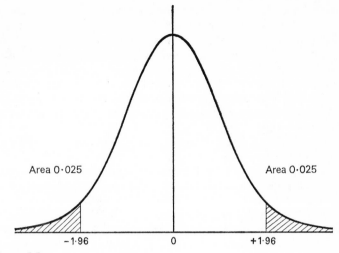

Figure 5.7

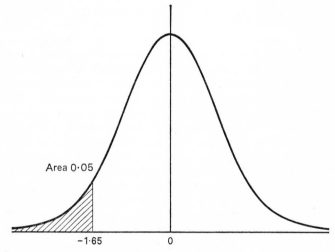

Figure 5.8

In Figure 5.8 the shaded area represents $0 \cdot 05$ of the total area. If, therefore, the z value calculated from the biscuit samples is $-1 \cdot 65$ or less, it will fall into the negative tail, and there will be a probability of less than $0 \cdot 05$ that the sample could have been drawn from a population (of all packets of this type of biscuit) in which the population mean weight is *at least* 225 g. What must be stressed is that the critical value for z corresponding to a probability of 0·05 is lower for a one-tailed test than it is for the corresponding two-tailed test, and in a sense it is 'easier' to obtain a statistically significant difference.

In some cases, the form of the data may determine the direction of any possible difference between the two values. For example, in recent years inflationary pressures have ensured that *money* wages have moved in one direction only—upwards. Consider the following information derived from family expenditure surveys. In 1966 a random sample of 228 full-time women employees in manufacturing industries had a mean weekly income of £10·03. In 1967, the mean weekly income of a random sample of 423 full-time women employees in manufacturing industries had risen to £10·65. Has there been an increase in the weekly money earnings of women employees in manufacturing industry, or is the apparent increase due to sampling fluctuations? Here, for a purely non-statistical reason—the general inflationary situation—the direction of the data is one way only, and a one-tailed test would be appropriate.*

Whether tests are two-tailed or one-tailed may be fortuitous, depending on the wording of the alternative hypothesis. Suppose that in a random sample of voters, 45% said that they would vote for Party A at the next general election. Party A, on the other hand, might claim it had the support of 48% of all voters. If this were subjected to a significance test, the null hypothesis would be that there was no real difference between the sample supporters of Party A (45%) and their support in the population as a whole (48%); the observed difference would be considered to be due to sampling error. The alternative hypothesis could be *either* that there was a difference, and that on the basis of the sample, support for Party A in the population of voters was not 48% (a two-tailed test), *or* that fewer than 48% of the electorate supported Party A (a one-tailed test). The difference depends on the form of words, but leads to a different critical value of z being used for test purposes, at a given probability level.

In effect, the two-tailed test is more 'stringent' than the one-tailed test, since it involves higher critical values for z for a given probability. This being so, where any doubts exists as to the direction of the data, or to the correctness of a one-tailed

---

* This would not be the case if what was at issue was *real*, as opposed to *money* wages, since although money wages have increased, real wages may have declined. The direction of possible change would no longer be determined by the situation.

test, a two-tailed test should be used. Readers should not forget that the results of two-tailed tests and one-tailed tests are not strictly comparable because of the different critical values involved. So far in this chapter, two-tailed tests have been used, partly to avoid confusion, and partly because of the more rigorous nature of the two-tailed test.

If comparability is required, then one-tailed tests can be conducted at the same critical values as for two-tailed tests. What is then altered is the probability which is, of course, halved. Thus in a one-tailed test, a critical value of z of $1 \cdot 96$ is equivalent to $p = 0 \cdot 025$, $2 \cdot 58$ corresponds to $p = 0 \cdot 005$, etc.

Examples of one-tailed tests are given in the exercises at the end of the chapter.

5.11 The interpretation of significant differences

This section provides a few words of caution to the users of significance tests. The purpose of significance tests is to draw verbal conclusions about research or survey material of some kind, arising from the use of numerical tests. The rejection of the null hypothesis implies a difference between the two values, the subject of the test, which cannot be attributed to chance, i.e. to sampling error. Such differences are 'statistically significant'. In many projects, which involve testing for significant differences, a whole series of such tests is carried out. Where this is so, depending on the probability level used for testing, a small number of significant differences must be expected to arise by chance. For example, at $0 \cdot 05$ probability, in a battery of 100 tests, about five tests should by chance throw up apparently significant differences, even though none really exist. This clearly leads to a most important reservation; the single significant difference, unsupported by other tests or other evidence, may well be maverick, and cannot be taken to prove something inconsistent with the body of evidence from other test results. After all, sooner or later in a battery of tests, significant differences will occur by *chance*. Where this may be the case (i.e. a small number of significant differences in a large number of related test situations), great caution should be taken in interpreting the result, and supporting tests or other evidence should be sought.

When a large proportion of the tests on a body of related data result in significant differences, then clearly the results of one test are consistent with results of another, and the statistical conclusions from the tests (of the existence of statistically significant differences) are supported.

What must be stressed is that significant differences are statistical in origin. They may indicate possibilities for further investigation; they may suggest relationships or association between variables, or 'cause and effect' patterns. They cannot of themselves prove the existence of relationships or establish a 'cause and effect' sequence. But they must *not* be looked at in isolation, or viewed as purely numerical results. The end product of any significance test is not simply '$|z| \geqslant z_{0.05}$ therefore reject the null hypothesis'. The end result is a verbal conclusion about the material subjected to the test, a conclusion which should involve non-statistical information about this basic material. What must never be lost from sight is that the whole purpose of a significance test is to draw meaningful conclusions about experimental, survey or other inquiry data.

Exercises

(Sections 5.2 to 5.5)

1. A random sample of 100 children born in 1960 and all now living in a particular area, had a mean height of 122 cm with a standard deviation of $2 \cdot 4$ cm. Could these children have been drawn from a population (of all children born in 1960) whose mean height is 120 cm? Test at $0 \cdot 01$ probability.

2. A group of 50 male students had a mean score of 117 with standard deviation 6.3, in a series of tests. Could this group of students be looked upon as a random sample drawn from a student population with a mean score of $118 \cdot 5$?

3. Calculate the mean age of the distribution of 1,731 female practitioners shown in Table 5.1, and also the standard deviation of the mean. If the mean age of all practitioners (male and female) is $48 \cdot 0$ years, could the mean age of female practitioners be considered to be significantly different from the age of all practitioners?

4. A random sample of 26 full-time men employees in East Anglia had mean weekly earnings of £23·40 with standard deviation £3·80. The mean weekly earnings of all full-time men employees in the UK was £26·00. Could the full-time men employees in East Anglia be looked upon as a random sample of all full-time men employees in the UK?

(Section 5.6)

5. The mean weekly wage of a random sample of 225 women in full-time office employment was £12·70 with standard

Table 5.1 *Number of female general practitioners in practice on 1 October, 1965 in the National Health Service*

Age group	Number
Under 35	207
35–39	356
40–44	304
45–49	223
50–54	177
55–59	163
60–64	189
65–69	84
70 & over	28

Source: *Report of the Ministry of Health, 1963.*

deviation of £1·90. A second random sample of 441 women employed full-time in manufacturing industry had a mean weekly wage of £12·15 with standard deviation of £1·20. Do these samples suggest a difference in the wages paid to women working full-time in offices and manufacturing industry respectively?

6. (*a*) From the data in Table 5.2, calculate the mean size of household, and standard deviation, for households in the North and East Midlands regions respectively.

(*b*) On the basis that the data in Table 5.2 are drawn from

Table 5.2 *Size of household 1968*

Number of persons	North region	East Midlands region
One	76	52
Two	133	141
Three	113	93
Four	88	65
Five	60	31
Six or more	30	18
Total	500	400

Source: Adapted from data given in *Family Expenditure Survey, 1968*

random samples, is there a statistically significant difference between the mean size of household for the two regions?

7. The mean height of a group of 525 male students was 181·00 cm with standard deviation 3·50 cm. Another group of 50 male students has a mean height of 180·00 cm and standard deviation of 3·80 cm. On the basis that both groups are random samples, consider whether there is a significant difference between the mean heights of the students in the two groups.

(Section 5.7)

8. In 1968, 29 households in the South West region rented furnished accommodation paying a mean weekly rent of £2·50 with standard deviation £0·82. In Wales, the mean weekly rent of 16 households in furnished accommodation was £2·06 with standard deviation £0·65. On the basis that these are random samples, is there a statistically significant difference between the mean weekly rents of furnished accommodation in the South West region and Wales?

(Section 5.8)

9. Figures for *all* children in primary schools in 1963 showed that in that year 65% of all primary school children were in classes of 31 pupils or more. In 1968, in a random sample of 480 primary school children, 67% were found to be in classes of 31 or more.

Does this suggest any change between 1963 and 1968 in the proportion of all primary school children in classes of 31 pupils or more?

10. An 'eve of the poll' survey of a sample of 1,739 voters found that 50·6% of the voters intended to vote for the Labour Party the following day. The actual result at the General Election was that 48·7% of the voters voted for the Labour Party.

Could the sample be looked upon as a random sample?

11. In a random sample of persons aged 65 years and over, 34% of a total of 197 single men and widowers and 49% of a total of 607 single women and widows were found to be solely dependent on state benefits for the whole of their incomes. For all persons aged 65 and over (single men, single women and couples), 37% had previously been found to be solely dependent on state benefits.

Consider whether the random samples of men and women respectively show significant differences in the percentages dependent on state benefits as compared with the established population percentage.

(Section 5.9)

12.   In a random sample of 600 parents in Town A, 366 expressed a preference for single-sex, as opposed to mixed secondary schools. Out of a similarly drawn random sample of 400 parents in Town B, 236 preferred single-sex secondary schools. Is there a statistically significant difference between the proportion of parents in the two towns who preferred single-sex secondary schools?

13.   (*a*) From the data in Table 5.3, do you consider that there is a significant difference between patients in teaching and other hospitals with regard to the proportion who were entirely enthusiastic in their view of the nurses?

Table 5.3 *Patients' views of nurses in teaching and other hospitals (excluding maternity patients)*

View of nurses	Teaching hospitals %	Other hospitals %
Entirely enthusiastic	66	54
Inter- mediate	23	25
Some criticism	11	21
	100	100
No. of patients	72	536

Source: Ann Cartwright, *Human Relations and Hospital Care,*
    (Routledge & Kegan Paul, 1964).

(*b*) As for (*a*) above, in regard to patients who voiced some criticism of the nurses.

14.   On the basis that the data in Table 5.4 is obtained from randomly drawn samples, test whether between 1966 and 1968 there has been a statistically significant change in (*a*) the pro-

portion of one-person households; and (*b*) the proportion of households consisting of six or more persons.

Table 5.4 *Family Expenditure Survey: size of households*

No. of persons	1966 % of households	1968 % of households
1	14·0	16·2
2	30·8	30·2
3	21·1	20·6
4	18·7	18·3
5	8·7	8·8
6 or more	6·7	5·9
	100·0	100·0
Total number of co-operating households	3,274	7,184

(Section 5.10)

15. A random sample of 239 women employed full-time in service industries had a mean weekly income of £11·32 with standard deviation £2·11. The mean weekly income of all women employed full-time was £12·20. Are women employed full-time in service industries paid less on average than all women full-time employees?

16. In the West Midlands region, a random sample of 316 households had a mean weekly income of £25·85 with standard deviation of £12·73. A similar random sample of 351 households in the Yorkshire and Humberside region had a mean weekly income of £23·14 with standard deviation of £13·35. Test the hypothesis that the mean weekly income of households in the West Midlands region is higher than that of households in the Yorkshire and Humberside region.

17. In a particular area, random samples of parents were asked, in successive years, for their preferences as between neighbourhood secondary schools and selective secondary schools. In the first year, 236 out of a random sample of 400 parents expressed a preference for neighbourhood secondary schools. In the following year, 357 out of a random sample of

500 parents preferred neighbourhood secondary schools. Does this suggest an increase in the support for neighbourhood secondary schools between the dates of the first and second surveys?

Further reading

General discussions of significant differences are to be found in:

A. Bradford Hill, *Principles of Medical Statistics* (The Lancet, 8th ed. 1966) chapters 11 and 12.
F. Conway, *Sampling, an Introduction for Social Scientists* (Allen and Unwin, 1967) chapter 12.
G. Kalton, *Introduction to Statistical Ideas for Social Scientists* (Chapman and Hall, 1966) chapter 4.
S. M. Dornbusch and C. F. Schmid, *A Primer of Social Statistics* (McGraw-Hill, 1955) chapters 14 and 15.
  Type I and Type II errors and the alternative hypothesis are dealt with in:
J. E. Freund and F. J. Williams, *Modern Business Statistics* (Pitman, 2nd ed. 1970) chapter 10.
H. T. Hayslett, *Statistics Made Simple* (W. H. Allen, 1968) chapter 7.
W. A. Wallis and H. V. Roberts, *Statistics: A New Approach* (Methuen, 1957) chapter 12.

The magnitude of statistically significant differences

6.1 Population mean and sample mean

So far, null hypotheses have been set up, and tests for statistically significant differences carried out on the basis of there being 'no difference' between the two values involved. It is hypothesized either that the sample has come from the known population, or that the two samples have come from the same population. Questions may arise in relation to the magnitude of an observed difference. If the difference between sample and population means is 'significant', from what population (in terms of its mean) may the sample have been drawn? Or for two sample means, how big is the difference between the two population means? The same questions can be asked again in relation to proportions.

On the basis of information contained in the 1968 Family Expenditure Survey, the mean size of a random sample of 7,184 households was 2·95 persons, with a standard deviation of 1·53 persons. According to the 1961 Census of Population, the mean size of household was then 3·04 persons. A significance test could be used to test whether the difference between the sample figure for 1968 and the population figure for 1961 could be due to chance, or whether the difference should be considered significant. On the basis of say, 0·01 probability, the critical value would be $z_{0·01} = 2·58$. The calculated value of z would be given by

$$z = \frac{\bar{x} - \mu}{s/\sqrt{n}} = \frac{2·95 - 3·04}{1·53/\sqrt{7184}} = \frac{-0·09}{0·0181} = -4·98.$$

Clearly, there is a significant difference between the two values, since the calculated $z = -4·98$ and therefore $|z| > z_{0·01} (= 2·58)$. It may not be enough to know that the apparent difference is 'real'. Since there is a difference, and the sample with mean size of household of 2·95 persons does not come from a population with mean size of household of 3·04 persons, the next step may be to consider whether the sample could have been drawn from a population with mean household size of, say 3·00 persons? The appropriate sampling distribution would be obtained by taking as the observed difference, the difference between the sample mean and the new hypothesized population mean:

$$z = \frac{\bar{x} - \mu'}{s/\sqrt{n}}$$

where μ' is the new hypothesized population mean. In this example, using the new population mean of 3·00,

$$z = \frac{2 \cdot 95 - 3 \cdot 00}{0 \cdot 0181} = \frac{-0 \cdot 05}{0 \cdot 0181} = -2 \cdot 78.$$

Again, at $-2 \cdot 78$, $|z| > z_{0 \cdot 01}$, so that, once more, the null hypothesis is rejected and the conclusion reached that the sample is unlikely ($p < 0 \cdot 01$) to have been drawn from a population with mean 3·00 persons per household. The corollary is, therefore, that the sample is likely to have come from a population with mean size of household less than 3·00 persons.

The problem can be approached another way. A possible population mean can be found by first determining the probable error that is acceptable on this population estimate. Such a population mean is effectively at one extreme of the appropriate confidence limits.

In this particular example, the magnitude of the z value initially calculated suggests very strongly that there has been a real decrease in the mean size of household. Given a value of $z = -4 \cdot 98$, as above, the probability that the size of household has increased since 1961 is extremely small (about 3 in 10,000,000) so that it is quite reasonable to assume a real decline in the mean size of household between the two dates. The decline in the population mean size of household may not, however, be as great as the difference between the previous known population mean and the sample mean from the 1968 Family Expenditure Survey data. Part, at least, of the decline may be apparent, not real, and due to sampling error. How much greater than the sample mean might the population mean be? Continuing with the same test level, there is a 0·99 probability that the true population mean would differ from the sample mean by not more than $2 \cdot 58 \times se_{\bar{x}}$; that is, by $2 \cdot 58 \times 0 \cdot 0181$ persons, or 0·0467 persons.

If what is at issue is the minimum size of household, then this should be treated as a one-tailed situation (see section 5.10). Thus, retaining the same critical value of z indicates either a probability of 0·99 that the true population mean in 1968 was not greater than 2·9967 persons, or less than 2·9033 persons; or, alternatively, a probability of 0·995 that the true population mean was not greater than 2·9967 persons. The different situations are shown in Figure 6.1.

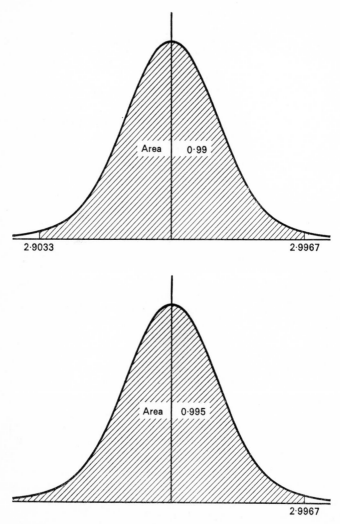

Figure 6.1

In this particular example, therefore, there is a $0 \cdot 995$ probability that the mean size of household in 1968 was not greater than $2 \cdot 9967$ persons, so that there would have been a real decline in mean household size of at least $0 \cdot 0433$ persons since 1961.

The calculation of the change in the population mean can be done by determining the change in relation to the magnitude of the acceptable error, itself dependent on the probability selected. On the basis of a one-tailed situation and taking probability $0 \cdot 995$ (equivalent to $0 \cdot 99$ for a two-tailed situation), let I be the change in household size.

Then $I = \mu' - \mu$

and* $\mu' = \bar{x} + 2 \cdot 58 \times \mathrm{se}_{\bar{x}}$

Thus $I_{0 \cdot 995} = \bar{x} + 2 \cdot 58 \times \mathrm{se}_{\bar{x}} - \mu$

$\qquad = 2 \cdot 95 + 0 \cdot 0467 - 3 \cdot 04$

$\qquad = 2 \cdot 9967 - 3 \cdot 04$

$\qquad = -0 \cdot 0433.$

The result is the same, whichever way the situation is approached: there is a $0 \cdot 995$ probability that there has been a real decline in the mean size of household of not less than $0 \cdot 0433$ persons between 1961 and 1968. Thus the $0 \cdot 995$ probable decline in household size between 1961 and 1968 is $0 \cdot 0433$ persons. The $0 \cdot 995$ probability represents the probability that the decline in household size is at least as great as $0 \cdot 0433$ persons.

What has been considered so far is the change indicated by an observed difference between a sample mean and a population mean. The same arguments can be developed to look at the difference between two sample means, drawn from different populations, and thereby to find out something of the difference between the two population means.

6.2   The difference between sample means

Consider the Family Expenditure Survey data, mentioned in section 5.10 above, about random samples of women employed full time in manufacturing industry. In 1966, the mean weekly income of 228 women was £10·03 and the standard deviation

---

* The single estimate of the population mean is given here by ($\bar{x} + 2 \cdot 58 \times \mathrm{se}_{\bar{x}}$) since the one-tailed position is for a population mean greater than the sample mean. If this were reversed, then in a situation in which the possible population mean was less than the sample mean, the estimate of the population mean would be based on ($\bar{x} - 2 \cdot 58 \times \mathrm{se}_{\bar{x}}$).

£2·423; in 1967, the mean weekly income of 423 women was £10·65 and the standard deviation £2·395. A straight null-hypothesis significance test, at 0·05 probability, on the basis that the samples were drawn from the same or similar populations (i.e. that there had been no change in their money earnings between 1966 and 1967), would lead to a calculated z value as follows:

$$z = (\bar{x}_1 - \bar{x}_2)\Big/\sqrt{\left(\frac{s_1{}^2}{n_1} + \frac{s_2{}^2}{n_2}\right)}$$

$$= (10\cdot65 - 10\cdot03)\Big/\sqrt{\left(\frac{2\cdot423^2}{228} + \frac{2\cdot395^2}{423}\right)}$$

$$= \frac{0\cdot62}{0\cdot198} = 3\cdot13.$$

At $z = 3\cdot13$, $|z| > z_{0\cdot05}$ ($=1\cdot96$); clearly there is a significant difference between the earnings of women employees in manufacturing industry between 1966 and 1967. The direction of change (on non-statistical as well as statistical grounds) is clearly in the direction of an increase in money income. How big might the increase be? Suppose the claim is that the real increase in money earnings has been not less than £0·20 a week. What would have to be considered here would be the theoretical sampling distribution of the difference between sample means, drawn from different populations, and with a given difference between the population means. Such a theoretical distribution is a normal distribution with mean of ($\mu_1 - \mu_2$) and standard deviation ($se_{\bar{x}_1 - \bar{x}_2}$) of

$$\sqrt{\left(\frac{s_1{}^2}{n_1} + \frac{s_2{}^2}{n_2}\right)}.$$

The null hypothesis would be that the change in money earnings had not exceeded £0·20 a week. The z value would be calculated on the basis of the excess of the observed difference between sample means over and above the hypothesized difference between the population means. Thus

$$z = \{(\bar{x}_1 - \bar{x}_2) - (\mu_1 - \mu_2)\}\Big/\sqrt{\left(\frac{s_1{}^2}{n_1} + \frac{s_2{}^2}{n_2}\right)}.$$

Suppose that in this example the test is again being conducted at 0·05 probability, so that the null hypothesis will be rejected if $|z| \geqslant z_{0\cdot05}$ ($=1\cdot96$).

$$z = \{(10\cdot65 - 10\cdot03) - 0\cdot20\}\Big/\sqrt{\left(\frac{2\cdot423^2}{228} + \frac{2\cdot395^2}{423}\right)}$$

$$= (0 \cdot 62 - 0 \cdot 20)/0 \cdot 198 = 0 \cdot 42/0 \cdot 198$$

$$= 2 \cdot 12$$

$|z| > z_{0 \cdot 05}$ so that the null hypothesis is rejected. This means that there is a not less than $0 \cdot 95$ probability that the change in money earnings has varied from the difference between the sample means by at least £$0 \cdot 20$ a week. Variation here means variation in either direction – an increase or a decrease in money income, in relation to the difference between the sample means, greater than £$0 \cdot 20$. Since what is at issue is the minimum probable increase in income, this can be interpreted on the basis of a one-tailed situation, that there is a probability of $0 \cdot 975$ that the increase in money earnings is at least as great as £$0 \cdot 20$ a week.

Again, the difference between the two sample means can be used to obtain the $0 \cdot 95$ probable change in earnings between 1966 and 1967. Effectively, what is done is to adjust the difference between the sample means by the $0 \cdot 95$ probable error. This adjustment can be in either direction. That is, the difference between the sample means is both increased and reduced by the $0 \cdot 95$ probable error. For these data the $0 \cdot 95$ probable error is given by

$$1 \cdot 96 \times se_{\bar{x}_1 - \bar{x}_2} = 1 \cdot 96 \times £0 \cdot 198 = £0 \cdot 388.$$

There is, therefore, a $0 \cdot 95$ probability that the true change in weekly earnings $(\mu_1 - \mu_2)$ lies between $(\bar{x}_1 - \bar{x}_2) \pm$ the $0 \cdot 95$ probable error. That is, the true change lies between

$$(10 \cdot 65 - 10 \cdot 03) - 0 \cdot 388 \text{ and } (10 \cdot 65 - 10 \cdot 03) + 0 \cdot 388$$

that is $0 \cdot 62 - 0 \cdot 388$ and $0 \cdot 62 + 0 \cdot 388$

that is, between $0 \cdot 232$ and $1 \cdot 008$.

With the minimum increase of weekly income at issue, this is strictly a one-tailed situation. The lower limit alone is relevant. The result can be expressed by saying that where I is the increase in income, its minimum value (at $0 \cdot 975$ probability) is given by:

$$I_{0 \cdot 975} = (\bar{x}_1 - \bar{x}_2) - 1 \cdot 96 \times se_{\bar{x}_1 - \bar{x}_2}$$

$$= (10 \cdot 65 - 10 \cdot 03) - 0 \cdot 388$$

$$= 0 \cdot 232.$$

There is, therefore, a $0 \cdot 975$ probability that the actual increase in weekly income is at least £$0 \cdot 232$.*

6.3   Population proportion and sample proportion

Sample proportions and population proportions can be treated in exactly the same way. In the example in section 5.8 above, about the proportion of male students in a student population, the significance test produced a z value of $2 \cdot 66$. As the test was conducted at $0 \cdot 01$ probability, this indicated a significant difference, that is, the sample could not have come from a population in which the proportion of male students was $0 \cdot 70$. What might be the proportion of males in the population from which the sample students were drawn? Making use of probable errors, the $0 \cdot 99$ probable error in relation to the sample proportion of $0 \cdot 53$ in a sample of 60 students is

$$\pm \, 2 \cdot 58 \times se_p$$
$$= \pm \, 2 \cdot 58 \sqrt{(0 \cdot 53 \times 0 \cdot 47/60)}$$
$$= \pm \, 2 \cdot 58 \times 0 \cdot 064$$
$$= \pm \, 0 \cdot 165.$$

There is, therefore, a $0 \cdot 99$ probability that the sample, proportion $0 \cdot 53$, is drawn from a population whose proportion does not differ by more than $0 \cdot 165$ from the sample proportion. That is, there is a $0 \cdot 99$ probability that the population proportion is not less than $0 \cdot 365$ ($=0 \cdot 53 - 0 \cdot 165$) or greater than $0 \cdot 695$ ($=0 \cdot 53 + 0 \cdot 165$).

If the proportion of males in the student population had previously been established at $0 \cdot 70$, what change is indicated by a sample proportion of $0 \cdot 53$? Using the equivalent of $0 \cdot 99$ probability, the extent of the minimum decline in the proportion of male students can be obtained. Suppose that I is the actual change from a population proportion of $0 \cdot 70$; $\theta = 0 \cdot 70$, and the new population proportion is θ', the proportion in the population from which the sample is drawn. As before, this is strictly equivalent to a one-tailed test, since the *minimum* change in θ is sought.

---

* If $(x_1 - x_2)$ is negative, the minimum *decline* in μ_1 to μ_2 would be given by:
$$I_{0 \cdot 975} = (\bar{x}_1 - \bar{x}_2) + 1 \cdot 96 \times se_{\bar{x}_1 - \bar{x}_2}.$$
In both cases, the observed difference must be reduced by the appropriate probable error.

$$I = \theta' - \theta$$
and $\theta' = p + 2 \cdot 58 \times \mathrm{se}_p$
$$
\begin{aligned}
I_{0.995} &= p + 2 \cdot 58 \times \mathrm{se}_p - \theta \\
&= 0 \cdot 53 + 0 \cdot 165 - 0 \cdot 70 \\
&= 0 \cdot 695 - 0 \cdot 70 \\
&= -0 \cdot 005
\end{aligned}
$$

This indicates only a very minimal change in the population proportion. What must be stressed is that at $0 \cdot 995$ probability, the exercise indicates a real decline in the proportion of male students of at least $0 \cdot 005$, or one half of one per cent. A $0 \cdot 995$ probability that the decline is not less than this amount is also a $0 \cdot 995$ probability that the decline is $0 \cdot 005$ or more.

Had a less stringent probability been used, say $0 \cdot 975$, the decline shown by the sample would have been greater—a decline of $0 \cdot 041$ or $4 \cdot 1 \%$, indicating that there was a $0 \cdot 975$ probability that the sample was drawn from a population with a proportion of male students not exceeding $0 \cdot 659$ or $65 \cdot 9 \%$.

6.4 The difference between sample proportions

Suppose that two samples of students are drawn from two colleges, and that of the first sample of 400 students, 332 are male, and of the second sample of 600 students, 396 are male. What difference, if any, do the samples indicate, in the proportion of males at the two colleges?

A straightforward significance test can be carried out, based on the null hypothesis that there is no difference between the proportions of male students at the two colleges, and that any observed difference is due to sampling error. Using, for example, $0 \cdot 05$ probability, the calculated z value would be compared with $z_{0.05}$ ($=1 \cdot 96$), and the null hypothesis rejected if $|z| \geqslant z_{0.05}$.

Here, $z = \dfrac{p_1 - p_2}{\mathrm{se}_{p_1 - p_2}}$

and* $\mathrm{se}_{p_1 - p_2} = \sqrt{\left\{ p(1 - p)\left(\dfrac{1}{n_1} + \dfrac{1}{n_2}\right) \right\}}$

* This is the form of $\mathrm{se}_{p_1 - p_2}$ used when the two samples are hypothesized to have been drawn from the same population, and is based on the 'pooled estimate' for p. This is obtained by combining the results of the two samples, so that

$p = \dfrac{x_1 + x_2}{n_1 + n_2}$ or $\dfrac{n_1 p_1 + n_2 p_2}{n_1 + n_2}$

where x_1, x_2 are the sample numbers equivalent to p_1, p_2. That is, $p_1 = x_1/n_1$ and $p_2 = x_2/n_2$.

$$p_1 = 332/400 = 0\cdot83; \text{ and } p_2 = 396/600 = 0\cdot66$$

$$p = \frac{332 + 396}{400 + 600} = \frac{728}{1000} = 0\cdot728$$

$$se_{p_1 - p_2} = \sqrt{\left\{0\cdot728 \times 0\cdot272 \left(\frac{1}{400} + \frac{1}{600}\right)\right\}}$$

$$= \sqrt{0\cdot0008251}$$

$$= 0\cdot0287.$$

Thus, $z = \dfrac{0\cdot83 - 0\cdot66}{0\cdot0287} = \dfrac{0\cdot17}{0\cdot0287} = 5\cdot92.$

Since $z > z_{0\cdot05}$ ($=1\cdot96$), the null hypothesis is rejected.

Further, the z value is so high that there is an extremely small probability of the observed difference between the two sample proportions being due to sampling error ($p < 0\cdot0000005$). The next question, therefore, is how big might the difference between the proportions of male students in the two college populations be?

The examination of the magnitude of the difference between two sample proportions implies that the samples are in fact drawn from different populations. The standard error of the difference between two sample proportions from different populations is given by*

$$se_{p_1 - p} = \sqrt{\left\{\frac{p_1(1 - p_1)}{n_1} + \frac{p_2(1 - p_2)}{n_2}\right\}}.$$

If, as for the sample means in section 6.2 above, it is hypothesized that the difference between two population proportions is a given amount, say $(\theta_1 - \theta_2) = 0\cdot10$, or 10%, the significance test is then based on a null hypothesis that there is no difference between the two population proportions over and above $0\cdot10$, and that any observed difference in excess of that amount is due sampling error. The observed difference would be reduced by the hypothesized value of $(\theta_1 - \theta_2)$, here $0\cdot10$, and the z value would be calculated from:

$$z = \frac{(p_1 - p_2) - (\theta_1 - \theta_2)}{se_{p_1 - p_2}}$$

* The 'pooled estimate' of the population proportion used in the significance test above would not apply here, since the basic assumption is that the two samples are in fact drawn from two separate and distinct populations; p_1 and p_2 are being used as the 'best estimates' of θ_1 and θ_2, the proportions of male students in the populations from which the samples are drawn. See section 5.9 above.

$$se_{p_1 - p_2} = \sqrt{\left(\frac{0\cdot83 \times 0\cdot17}{400} + \frac{0\cdot66 \times 0\cdot34}{600}\right)}$$

$$= \sqrt{0\cdot000627}$$

$$= 0\cdot0250.$$

Thus $z = \dfrac{(0\cdot83 - 0\cdot66) - 0\cdot10}{0\cdot0250}$

$$= \frac{0\cdot17 - 0\cdot10}{0\cdot0250}$$

$$= \frac{0\cdot07}{0\cdot0250}$$

$$= 2\cdot80.$$

With $z = 2\cdot80$, $|z| > z_{0\cdot05}$, so that the null hypothesis is rejected. There is, therefore, a less than $0\cdot05$ probability that the actual difference between the population proportions is $0\cdot10$, and (since the z value is positive) a probability of $0\cdot95$ that the difference is greater than $0\cdot10$ (see Figure 6.2).

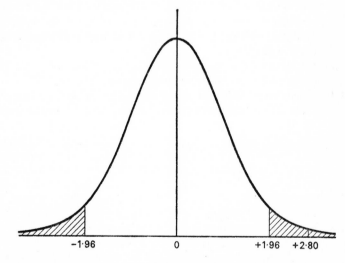

Figure 6.2

If this is treated as a one-tailed test situation, the result can be expressed by saying that there is a $0\cdot975$ probability that the difference between the proportions of male students in the two populations is greater than $0\cdot10$ or 10% (see Figure 6.3).

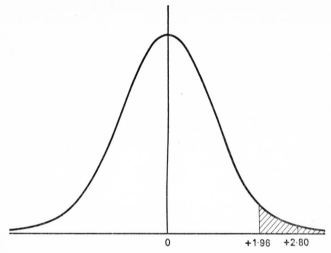

Figure 6.3

Had the calculated z value been less than the critical test value, then the null hypothesis could not have been rejected. A low z value would suggest that the two samples might well come from populations whose proportions differed by $0·10$. A high z value, but one below the critical value, would suggest an indeterminate situation.

The problem of the magnitude of the difference between the two population proportions can again be approached by finding the $0·95$ probable value of this difference. The $0·95$ probable error of the difference between the sample proportions is given by:

$$0·95 \text{ probable error} = 1·96 \times \text{se}_{p_1 - p_2}$$
$$= 1·96 \times 0·0250$$
$$= 0·049.$$

This is used to reduce the difference between the sample proportions, so that there is a $0·95$ probability that the difference between the sample proportions lies between

$(p_1 - p_2) \pm 0·95$ probable error

or $0·17 \pm 0·049$

that is, between $0·121$ and $0·219$.

But again, if what is wanted is one extreme position only, to give a minimum probable difference between the two population proportions, this becomes a one-tailed situation. Thus if the actual difference between the population proportions is represented by I,

$$I_{0.975} = (p_1 - p_2) - 0.95 \text{ probable error}$$
$$= 0.17 - 0.049$$
$$= 0.121.$$

This indicates that there is a 0.975 probability that the difference between the two population proportions is at least 0.121 or 12.1%.

6.5 Note on methods

Two general points should be made about all the situations examined in this chapter. First, minimum differences between two sample statistics have been investigated. Equally, maximum differences could be estimated. This would involve a similar argument, with a one-tailed situation at the opposite extreme. In the last example quoted above, there is a 0.975 probability that the difference between the two population proportions is at least 0.121. At the opposite extreme, obtained from the expression:

$$I_{0.975} = (p_1 - p_2) + 0.95 \text{ probable error,}$$

there is a 0.975 probability that the difference between the two population proportions does not exceed 0.219 or 21.9%. Users of significance tests who are concerned with the magnitude of differences between sample statistics must be careful to establish their test situations appropriately and to draw their conclusions correctly.

The second point concerns the levels of probability used in significance tests. Different probabilities will give rise to varying conclusions about the magnitude of actual differences between population parameters. The analogy here is with confidence limits. Inevitably, choice of probability must be somewhat arbitrary, depending upon the data which are being examined.

Exercises

(Section 6.1)

1. A random sample of 388 full-time men employees in manufacturing industry in the North region had mean weekly earnings of £22·52 with standard deviation £5·95. The mean weekly earnings of all full-time men employees in manufactur-

ing industry was £24·44. Is it likely that the weekly earnings of full-time men employees in manufacturing industry in the North region was less than the national figure by as much as £1·00 per week?

2.   In 1961, the mean age of head of household of all households in the UK was 54·5 years. A later random sample of 6,136 heads of household had a mean age of 55·2 years with standard deviation of 16·4 years. What is the 0·975 probable minimum increase in the mean age of head of household?

3.   In a particular large town in 1968, the mean length of time spent by all families in temporary welfare accommodation before being suitably rehoused was 9·3 months. In 1970, the mean length of time spent in temporary accommodation by a random sample of 83 families was 10·7 months with standard deviation 1·2 months. Using a 0·995 probability, what is the minimum increase in the length of time spent in temporary accommodation?

(Section 6.2)

4.   A random sample of 1,231 men manual employees engaged full-time in manufacturing industry had mean weekly earnings of £23·35 with standard deviation £3·66. A similarly drawn random sample of 477 full-time men manual employees in transport had mean weekly earnings of £22·62 and standard deviation of £3·91. What is the 0·975 probable minimum difference in earnings between full-time men manual employees in manufacturing industry and transport? What is the 0·975 probable maximum difference in earnings?

5.   The heights of random samples of 11-year-old girls in two areas were measured. In the first area, the mean height of 326 girls was 147·50 cm with standard deviation 2·80 cm. In the second area, the mean height of 453 girls was 148·70 cm with standard deviation 3·20 cm. Is there a statistically significant difference between the mean heights of 11-year-old girls in the two areas? If so, what is the 0·995 probable minimum difference between the mean height of all 11-year-old girls in the two areas?

(Section 6.3)

6.   A survey of all universities established that 8·9% of all students dropped out at the end of their first year. In a later random sample of 977 students in science departments, 109 dropped out at the end of their first year. By how much does the first-year drop-out rate of science students differ from the

previously established drop-out rate for all university students? Use 0·975 probability and estimate both maximum and minimum values.

7. In 1966, as a result of previous immunization, 52% of all children then under 16 years of age were regarded as protected against diphtheria. In 1970, a survey of 1,570 children under 16 years of age established that 58% were protected against diphtheria. On the basis that this was a random sample, estimate the 0·995 probable increase between 1966 and 1970 in the percentage of all children under 16 who were protected against diphtheria.

(Section 6.4)

8. In a random sample of 2,315 voters in a particular constituency, 1,097 were men and 1,218 were women. 52% of the men and 57% of the women said they supported the Conservative Party. Is there a statistically significant difference between men and women voters in this constituency in their support for the Conservative Party? What is the minimum percentage, at 0·975 probability, by which women voters who support the Conservative Party exceed men?

9. In a survey carried out in Town A, the proportion of households with television sets was 47% out of a random sample of 731 households. In a similar survey in Town B, 35% out of a random sample of 545 households had television sets. Could the difference between the percentages of all households with television sets in the two towns be as great as 5%?

10. Out of 1,450 first-year students at a particular university, 570 were studying courses in science departments and 880 in arts departments. In the science departments, 447 students were men and in arts departments 543 students were men. On the basis that these are random samples, what is the 0·975 probable difference between the percentages of men students in the science and arts departments respectively?

Tests for goodness of fit and association

7.1 Introduction

The preceding two chapters are concerned with tests involving a variable, that is, something that can be given a numerical value. Not all data come in the form of variables, and neither do all situations involving hypotheses lend themselves to a simple comparison between an actual value of a variable and a hypothesized value. It may be necessary to compare an experimental distribution with a corresponding series of values which would obtain under a particular hypothesis; alternatively, data may be categorized, rather than be in variable form. This latter becomes simply a form of 'head counting', analogous to tests involving proportions.

Consider, for example, the following data. A dice has been rolled 300 times, and these results obtained.

Face	No. of rolls	Face	No. of rolls
1	46	4	45
2	62	5	47
3	56	6	44

If the dice is balanced, each face would have an equal probability ($p = 1/6$) of appearing uppermost and, in 300 rolls, the mathematical expectation would be that each face would appear uppermost 50 times. Could these results be attributed to chance, that is, to sampling error? Or are the differences between the experimental results and the theoretical values such that they are unlikely to be due to chance? What arises fundamentally is a situation where a number of observed values are to be compared individually with others, these latter being based on some hypothesis. The problem resolves itself into one of 'goodness of fit' between the original data, and the values that would be expected under the terms of the hypothesis.

The situation can again be dealt with in terms of a null hypothesis, this time a null hypothesis of no difference between the two distributions of observed and expected results, any actual difference being due to chance or sampling error.

If the comparison was a simple one, between two situations, say of rolling a six or not rolling a six, the binomial distribution would be applicable. Rolling a six could be 'success', and the comparison would then be:

	Observed results	Expected results
Number of sixes	44	50
Other faces	256	250

This could be treated as a test of the difference between sample proportions.

But the binomial distribution (and the normal distribution as its approximation) cannot be used if more than two situations are involved. Then a multinomial distribution is needed, and finding a suitable continuous distribution as an approximation to what is strictly a discrete distribution is more complex than in the case of the binomial and normal distributions.

The basis for any test of 'goodness of fit' is random sampling, i.e. the observed values are themselves the result of random selection. Thus the test depends upon the characteristics of the sampling distribution that would arise if the set of values (here the number of times each face of the dice appears uppermost) is in fact the result of random sampling. In such circumstances, the differences between observed and expected values are attributable to chance, that is, to sampling error.

7.2 The χ^2 statistic and the chi-square distribution

The comparison between the observed and expected values is made by calculating the χ^2 statistic, where

$$\chi^2 = \sum_{i=1}^{k} \frac{(o_i - e_i)^2}{e_i}$$

and o_i and e_i are the corresponding observed and expected values, in a set which consists of k pairs of such values.

This statistic approximates to a chi-square distribution. But

I

since the χ^2 statistic only approaches the chi-square distribution as total frequency approaches infinity, the approximation strictly is only close for high observed and expected values, that is for large samples.

A chi-square distribution (like the normal distribution) is a continuous distribution, whereas the χ^2 statistic (like the binomial distribution) is discrete. The underlying assumption in relating the chi-square distribution and the χ^2 statistic is that individual differences between the observed and expected values behave as independent normal variates with a mean of zero. It implies that the mean difference between any pair of observed and expected values is zero. The χ^2 statistic is calculated from the sum of the squares of these independent normal variates in standard unit form (equivalent to z values). Comparison between the χ^2 statistic and a chi-square distribution provides a measure of the probability of the distribution of differences between the observed and expected values occurring by chance. The assumption, that the difference $(o_i - e_i)$ is normally distributed is itself only approximately true; the actual distributions of differences are discrete, not continuous.

A chi-square distribution is an asymmetrical continuous distribution which behaves as a relative frequency or probability distribution. It has a single 'tail', and is asymptotic to the horizontal axis. Very high values of chi-square may occur, but with very small frequency. All values are positive, since the differences are squared, unlike the normal curve which is

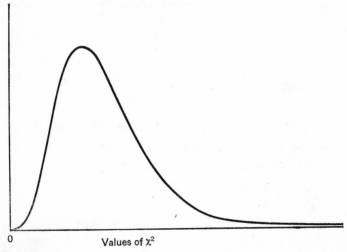

0   Values of χ^2

Figure 7.1

symmetrical round $z = 0$. Figure 7.1 shows a typical chi-square distribution.

The area under the curve represents the relative frequency with which specific values of chi-square occur. Thus, for a given value of chi-square the frequency with which that value, or larger ones occur is given by the area in the single tail of the distribution. As for the normal distribution, the total area under the curve is taken to be unity, and the proportion of area to the right of a chi-square value gives the relative frequency with which that value or higher ones would be expected to occur, in a large number of random samples. In sampling situations in which the χ^2 statistic has been calculated, the area to the right of its position along the horizontal axis of the chi-square distribution represents the probability of obtaining a particular value of χ^2 (or higher values) by chance, that is, as a result of sampling error.

7.3 Degrees of freedom

Degrees of freedom have been referred to previously in chapter 4 in connection with the Student-t distribution (see footnote on page 52). In comparing the t statistic with the theoretical distribution, degrees of freedom must be taken into account; the number of degrees of freedom reflects the number of ways in which two sets of data that are being compared are free to vary.

The χ^2 statistic is a cumulative statistic, and its value depends both on the magnitude of the difference between the observed and expected values, in relation to the expected value, and on the number of items contributing to χ^2. Other things being equal, χ^2 will be bigger, the larger the number of items in the distributions. This latter point has therefore to be taken into account, through the number of degrees of freedom inherent in a particular calculation of the χ^2 statistic. Once the χ^2 statistic has been calculated, it is compared with the chi-square distribution which is appropriate for that number of degrees of freedom. This is known as a chi-square test, and is analogous to the normal curve significance test. As for the Student-t distribution, there is a separate chi-square distribution for each number of degrees of freedom.

In a chi-square test the number of degrees of freedom depends upon the terms of the null hypothesis and the restrictions that are thereby imposed as part of the calculation of the expected values. In the example in section 7.1 above, about the results of 300 rolls of a dice, the hypothesis is that the dice is balanced. The null hypothesis therefore, is that there is no real difference

I*

between the experimental results, and those that would be expected under the terms of the hypothesis that the dice is balanced. Any observed difference is attributed to chance or sampling error. The expected values are therefore obtained by taking p = 1/6 for each face of the dice, and by taking the same total frequency for the observed and expected values. The first condition does not restrict variation between the two sets of observed and expected values, but the second does. After five expected values have been calculated, the sixth is determined by the condition that the separate sums of the observed and expected values shall be the same. One degree of freedom has been lost. The number of pairs of values is six; since one degree of freedom has been lost, the number of degrees of freedom left is $(6 - 1) = 5$.

In other forms of this test for goodness of fit, more degrees of freedom are lost. The calculation of the degrees of freedom will be discussed again as it arises.

7.4   The chi-square test

Reverting now to the data about the 300 rolls of a dice, the chi-square test is conducted as follows.

1.   The dice is hypothesized to be balanced. This is turned into a null hypothesis that there is no difference between the experimental results obtained in rolling a dice 300 times, and the results which would be expected in 300 rolls on the basis of the hypothesis (for each face, $p = 1/6$).

2.   The number of degrees of freedom is worked out. Here it is five.

3.   A suitable probability for testing is chosen, say $p = 0.05$.

4.   The criterion for rejecting the null hypothesis is established. The value in the chi-square table for five degrees of freedom, and corresponding to 0.05 probability is obtained. This is $\chi^2_{0.05} = 11.07$. Thus the null hypothesis will be rejected if

$$\chi^2 \geqslant \chi^2_{0.05} \; (=11.07)$$

and it will not be rejected if

$$\chi^2 < \chi^2_{0.05}$$

5. χ^2 is calculated as $\chi^2 = \sum_{i=1}^{k} \dfrac{(o_i - e_i)^2}{e_i}$.

In this example, the calculation is as follows:

Face	o	e	$o-e$*	$(o-e)^2$	$(o-e)^2/e$
1	46	50	-4	16	0·32
2	62	50	$+12$	144	2·88
3	56	50	$+6$	36	0·72
4	45	50	-5	25	0·50
5	47	50	-3	9	0·18
6	44	50	-6	36	0·72

$$\Sigma \frac{(o-e)^2}{e} = 5·32$$

6.   The χ^2 value as calculated is 5·32. This is less than 11·07. $\chi^2 < \chi^2_{0·05}$ so that the null hypothesis cannot be rejected. On the basis of the test criterion the differences between the observed and expected values could be attributed to chance, and

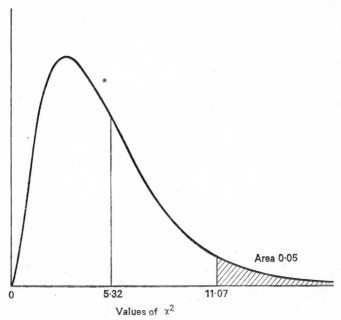

Values of χ^2

Figure 7.2

* The differences between the observed and expected values cancel out when regard is paid to the sign of the differences. That is, $\Sigma(o-e) = 0$. Since each value of $(o-e)$ is to be squared, the sign of the difference is not relevant, and could be ignored by taking $|o-e|$. However, there is a slight practical advantage in checking at this point that the sum of the differences is zero.

there is no evidence to suggest that the dice was not balanced. An examination of the table of chi-square distributions suggests that the probability of the calculated value of χ^2 occurring by chance is about 0·4. The original hypothesis, that the dice is balanced, is therefore accepted.

Figure 7.2 shows the position of the calculated value of χ^2 in relation to the chi-square distribution for five degrees of freedom.

7.5   The chi-square test as a test for normality

It is often claimed that data is 'normal' or 'approximately normal', meaning that a particular set of figures corresponds reasonably closely to a normal distribution. The chi-square test can be used to examine this hypothesis. Any set of figures alleged to be 'normal' can be compared with a normal distribution which has the same mean, standard deviation and total frequency as the observed data. Thus the chi-square test becomes a test for normality in data—a test of whether a particular distribution can be considered normal.

Suppose it is alleged that height is distributed normally in the adult population. The student population referred to in chapter 3 could be looked upon as a random sample drawn from the adult population. If so, then the variable, height should be distributed normally in the student population.

Table 7.1 sets out the frequencies from the student height distribution, together with the corresponding frequencies for a normal distribution with the same mean, standard deviation and total frequency as the student population.* Table 7.2 is the χ^2 calculation.

The steps in this use of the chi-square test are as follows.

1.   In this example, the distribution of student heights is hypothesized to be a normal distribution. The null hypothesis is, therefore, that there is no difference, other than that attributable to random sampling error, between the distribution of student heights and a normal distribution with the same mean, standard deviation and total frequency.

2.   The number of degrees of freedom remaining is given by the number of pairs of observed and expected frequencies minus the number of degrees of freedom lost in calculating the expected values. The distribution of expected values has the same mean, standard deviation and total frequency as the

---

* The method of calculating the corresponding expected frequencies is set out in a note at the end of this section.

Table 7.1   *Heights of 900 students*

Height cm	Number of students	
	Observed distribution	Expected distribution
160 and under 164	31	35·6
164 and under 168	122	88·3
168 and under 172	154	158·6
172 and under 176	175	207·6
176 and under 180	197	192·8
180 and under 184	136	128·3
184 and under 188	65	61·9
188 and under 192	16	21·7
192 and under 196	4	5·2
All heights	900	900·0

Table 7.2   *Heights of 900 students: the χ^2 calculation*

o	e	$o-e$	$(o-e)^2$	$(o-e)^2/e$
31	35·6	−4·6	21·16	0·59
122	88·3	+33·7	1135·69	12·86
154	158·6	−4·6	21·16	0·13
175	207·6	−32·6	1062·76	5·12
197	192·8	+4·2	17·64	0·09
136	128·3	+7·7	59·29	0·46
65	61·9	+3·1	9·61	0·16
16	21·7	−5·7	32·49	1·50
4	5·2	−1·2	1·44	0·28

$$\chi^2 = 21 \cdot 19$$

observed data so that three degrees of freedom have been lost.
There are nine pairs of values, so that six degrees of freedom
remain.

   3. A suitable probability for testing is chosen, say $p = 0 \cdot 01$.

   4. The criterion for rejecting the null hypothesis is established. For six degrees of freedom, $\chi^2{}_{0 \cdot 01} = 16 \cdot 81$. Therefore
the null hypothesis will be rejected if $\chi^2 \geqslant \chi^2{}_{0 \cdot 01}$ ($=16 \cdot 81$), and
not rejected if $\chi^2 < \chi^2{}_{0 \cdot 01}$.

5. The expected values are calculated in accordance with the null hypothesis.

6. χ^2 is calculated; in this sample $\chi^2 = 21 \cdot 19$.

7. The calculated value exceeds $\chi^2_{0 \cdot 01}$ ($=16 \cdot 81$). Therefore the null hypothesis is rejected. The distribution of student heights cannot be considered to approximate to a normal distribution. The actual calculated value of χ^2 is just less than $\chi^2_{0 \cdot 001}$ ($=22 \cdot 46$) showing that the probability of the difference between the set of observed and expected values being due to chance or sampling error is less than $0 \cdot 01$, but slightly in excess of $0 \cdot 001$, or one tenth of 1 per cent.

8. The conclusion is therefore *either* that height is not normally distributed in the adult population, *or* that the students cannot be looked upon as a random sample of the adult population. In support of the second alternative, there are a number of ways in which the students will differ from the total adult population, such as age, socio-economic grouping, and possibly proportion of each sex. Certainly more evidence would be required before concluding that because height was not normally distributed in this group of students, height was not normally distributed in the adult population in total.

Note on the calculation of expected normal frequencies

The actual calculation of the corresponding normal frequencies is somewhat tedious. First the mean and standard deviation of the observed distribution are required. These are given in section 3.2; the mean is $174 \cdot 64$ cm and the standard deviation $6 \cdot 86$ cm. The class boundaries, (1), are then turned into standard normal deviates, or z values, by (2), subtracting the mean from each class boundary and then, (3), dividing the difference by the standard deviation. By using a normal curve area table, (4), and (5), the fraction of total area falling between the class boundaries can be obtained. These fractional areas represent expected frequencies corresponding to the classes of the observed distribution. Finally, (6), the fractional areas are multiplied by a suitable factor to bring their total frequency to that of the observed distribution.

The calculation is shown in Table 7.3. The column numbers are the steps in the calculation referred to above.

7.6 The chi-square test and contingency tables

Basically, a chi-square test may be used in any circumstances in which a set of observed data can be compared with a corres-

Table 7.3 *Calculation of expected normal frequencies*

(1)	(2)	(3)	(4)	(5)	(6)
Class boundaries (cm)	Difference from mean (cm)	Divided by standard deviation	Area from 0 to z	Area between class boundaries	Multiply by 900 / 0·9852
159·50	−15·14	−2·21	0·4864		
				0·0390	35·6
163·50	−11·14	−1·62	0·4474		
				0·0966	88·3
167·50	−7·14	−1·04	0·3508		
				0·1736	158·6
171·50	−3·14	−0·46	0·1772		
				0·2272	207·6
175·50	0·86	0·13	0·0500		
				0·2111	192·8
179·50	4·86	0·71	0·2611		
				0·1404	128·3
183·50	8·86	1·29	0·4015		
				0·0678	61·9
187·50	12·86	1·87	0·4693		
				0·0238	21·7
191·50	16·86	2·46	0·4931		
				0·0057	5·2
195·50	20·86	3·04	0·4988		
		Total frequencies		0·9852	900·0

ponding set of theoretical values, the data being treated as arising from random samples. There is a certain type of data, somewhat different from that already discussed above, for which this test is particularly useful.

Circumstances often arise in which the same data is spread over two separate sets of categories or attributes. The question that arises is whether or not the place of a sample unit in one distribution (or set of attributes) is in any way related to its position in the other distribution. Is there in fact an association between attributes?

Referring to the student data used in preceding chapters, suppose that the 900 students were divided according to their academic departments and according to sex, so that the categorized data in Table 7.4 was obtained.

In Table 7.4, which is known as a contingency table, the students have been distributed according to two sets of attri-

Table 7.4   *Department and sex of 900 students*

Department	Men	Women	All students
Arts	165	185	350
Science	168	92	260
Economics/ social science	115	105	220
Music	32	38	70
All departments	480	420	900

butes—academic department and sex. Each unique pair of attributes within the contingency table is referred to as forming a cell, so that each cell contains the number of students possessing a particular pair of attributes one from each set. Are these attributes related, or are the students spread at random across the two distributions? Is a woman student more or less likely than a man to be in a particular department? Or are any apparent variations simply due to chance? If the students can be looked upon as a random sample, do the data in Table 7.4 suggest that there is any association between the attributes of academic department and sex? Or is any apparent association simply due to chance that is to sampling error?

A brief look at the actual data in the Table shows that although there are more men than women students overall, in the departments of arts and music, women exceed men. Conversely, in science and economics /social science, men outnumber women. The data certainly suggest that there may be an association between academic department and sex. But is it more than a suggestion? Could the apparent association simply be due to chance?

If there is 'no association' between the way the data are distributed across the two sets of attributes, then the proportions of men and women in each academic department should be the same, and equal to the overall proportions of men and women among the 900 students. Similarly, considering the data from the point of view of department, the proportion of students from the different departments should be the same among both men and women, and equal to the overall proportion of students in each department. This is the expected position on the assumption of 'no association'. It is equivalent to the null hypothesis assumption of 'no difference' between two sample means, that is, the mean difference itself is 0. As discussed in chapter 5, the purpose of the normal curve significance tests is to examine to what extent observed differences involving sample statistics could be attributed to chance. Similarly with a chi-square test in connection with a contingency table, the null hypothesis of no association can be used to derive a set of values in conformity with the null hypothesis. As for the previous examples of chi-square tests, what is then necessary is to examine the magnitude of the differences between the observed values (the original data) and the expected values (based on the null hypothesis of no association) to see whether or not the differences are such that they could reasonably be attributed to chance.

In the student data, the assumptions are that the proportions of men and women students are the same in each department, and that the proportions in different departments are the same for both sexes. Since 480 out of 900 students are men, the overall proportion of men students is $0 \cdot 533$ or $53 \cdot 3\%$. In each department, $53 \cdot 3\%$ of students should be men. Similarly, since $46 \cdot 7\%$ of students are women, this should be the proportion of women in each department. Applying these percentages to the total for each department, a set of expected values is obtained. These are shown in the contingency table in Table 7.5 in brackets next to the observed value to which each relates.

Table 7.5 *Department and sex of 900 students—observed values and expected values*

Department	Men	Women	All students
Arts	165 (186·7)	185 (163·3)	350
Science	168 (138·7)	92 (121·3)	260
Economics/ social science	115 (117·3)	105 (102·7)	220
Music	32 (37·3)	38 (32·7)	70
All departments	480	420	900

The assumption of overall proportions applied to the original data produces a pattern of expected values which conforms to the original sub-totals as well as to the grand total of all students. Thus a number of degrees of freedom are lost by the restrictions placed on the calculation of the expected values. Effectively, the last expected value in each row or column is determined once the previous expected values in that row or column have been calculated. For a contingency table with r rows and k columns, the number of degrees of freedom remaining is given by $(r - 1) \times (k - 1)$. Rows and columns here refer to the body of the contingency table containing the observed and expected values; the sub-totals are excluded. In the student example above, there are four rows and two columns. The number of degrees of freedom is, therefore, $(4 - 1) \times (2 - 1) = 3 \times 1 = 3$.

It is necessary to make separate calculations of expected values equal in number to the remaining degrees of freedom, provided this is done by omitting the values for a marginal row and column. No value must be calculated which can be inferred from a previously calculated expected value. If, in this example, the expected numbers of men students in three departments (say arts, science, and economics/social science) are calculated, the remaining expected values can be obtained by subtraction from the sub-totals.

In practice, the simplest way to obtain any expected value is from the expression:

$$\text{expected value} = \frac{\textit{product of corresponding marginal sub-totals}}{\textit{grand total of all items}}$$

For the men in the arts department, the corresponding marginal sub-totals are the two which include the 186 men in

Table 7.6   *Calculation of expected value*

Department	Men	Women	All students
Arts	Men in arts⎯⎯⎯⎯→ department		Marginal sub-total (350)
Science			
Economics/ social science			
Music	↓	↘	
	Marginal sub-total (480)		Grand total (900)

the arts department. They are, therefore, 350 students (men and women together) in the arts department, and 480 men students (in all departments). The grand total is the overall total of 900 students. This is shown in Table 7.6.

The expected value is therefore

$$\frac{350 \times 480}{900} = 186 \cdot 7$$

This is precisely the same calculation as that already carried out above, since the fraction 480/900 is the overall proportion of men, which has been applied to the number of students in the arts department, 350, to give the expected value for that cell in the contingency table.

A most important point to appreciate is that under no circumstances can an observed value (e.g. actual number of men in the arts department) enter into the calculation of any expected value.

Once the expected values, equal in number to the degrees of freedom, have been obtained and inserted in the contingency table, the remaining expected values are arrived at by difference, as shown in Table 7.7. The marginal sub-totals, which must be the same for both observed and expected values, are $n_1, n_2, \ldots, n_6$, and the grand total is N.

Table 7.7 *Expected values*

Department	Men	Women	All students
Arts	e_1	$e_5 = n_1 - e_1$	n_1
Science	e_2	$e_6 = n_2 - e_2$	n_2
Economics/ social science	e_3	$e_7 = n_3 - e_3$	n_3
Music	$e_4 = n_5 - (e_1 + e_2 + e_3)$	$e_8 = n_4 - e_4$	n_4
	n_5	n_6	N

A simple cross-check can be made on the arithmetic, since e_8, the expected value in the last cell, can be calculated in two ways. It will be the same whether it is derived from the other values in the row or in the column in which it is situated. Thus

$$e_8 = n_4 - e_4; \text{ and } e_8 = n_6 - (e_5 + e_6 + e_7).$$

At this point in an actual calculation, it is worth checking that expected values do correctly cross-check in both directions to the marginal sub-totals.

Having obtained the expected values, which should be inserted in the contingency table as shown above, the calculation of χ^2 can be carried out as shown in Table 7.8.*

Table 7.8 χ^2 *calculation relating to Table 7.7*

o	e	$o-e$	$(o-e)^2$	$(o-e)^2/e$
165	186·7	−21·7	470·90	2·52
168	138·7	+29·3	858·50	6·19
115	117·3	−2·3	5·29	0·05
32	37·3	−5·3	28·09	0·75
185	163·3	+21·7	470·90	2·88
92	121·3	−29·3	858·50	7·08
105	102·7	+2·3	5·29	0·05
38	32·7	+5·3	28·09	0·86
				20·38

Suppose that in this case, the chi-square test is being carried out at $p = 0.05$. For three degrees of freedom, $\chi^2_{0.05} = 9.84$; the calculated value is 20·38. Thus $\chi^2 > \chi^2_{0.05}$, and the null hypothesis of 'no association' is rejected. The calculated value of χ^2 is in fact very large, exceeding $\chi^2_{0.001}$ (=16·27), so that the null hypothesis would be rejected also at $p = 0.01$ and $p = 0.001$. ~~accepted~~

If the null hypothesis of 'no association' is rejected, then the result is to assert positively that there is an association between the two sets of attributes, here sex and department. The chi-square test lacks direction; it does not indicate anything about the nature of the association, but simply provides evidence that the differences between the observed and expected values are too great to be attributable to chance, or sampling error. The nature of the association can be derived only by examining the relationship between the observed and expected values. Here,

---

* As before, the sum of the differences between the observed and expected values is zero. Further, the sum of these differences across any row, or down any column, also is zero. Where in one direction there are two attributes only (as here, men and women), differences between the observed and expected values are arithmetically the same, but with opposite signs. These points can be seen clearly in the third column of Table 7.8.

there is an association between lower than expected proportions of men in the departments of arts, economics/social science and music, with a higher than expected proportion in the department of science. Thus, on this basis, men are more likely than women to study science; women are more likely than men to study arts, economics/social science or music.

The test method, in relation to contingency tables is, therefore, as follows:

1.   The null hypothesis of 'no association' is established; that is, there is no association between the way in which the sample units are distributed across each of the two sets of attributes (sex and department). Thus it is hypothesized that there should be no difference, other than that which can be attributed to sampling error, between the observed values in the contingency tables and those values which would be expected on the basis of the null hypothesis.

2.   The number of degrees of freedom is worked out. For a contingency table of $r \times k$ cells, this will be $(r - 1) \times (k - 1)$, in this example three degrees of freedom.

3.   A suitable probability for testing is chosen, such as $p = 0.05$.

4.   The criterion for rejecting the null hypothesis is established. For three degrees of freedom, $\chi^2_{0.05} = 9.84$. Therefore the null hypothesis will be rejected if $\chi^2 \geqslant \chi^2_{0.05}$ $(=9.84)$ and not rejected if $\chi^2 < \chi^2_{0.05}$.

5.   Expected values equal in number to the degrees of freedom are calculated separately on the basis of the null hypothesis of no association, that is, on the assumption of applying overall proportions from one set of attributes to the other. In practice, the necessary expected values are obtained from the expression:

$$\text{expected value} = \frac{\textit{product of corresponding marginal sub-totals}}{\textit{grand total of all items}}.$$

The remaining expected values are obtained by difference from sub-totals.

6. χ^2 is calculated, as set out above. In this calculation $\chi^2 = 20.38$.

7.   In accordance with the critical value established in 4 above, since $\chi^2 > \chi^2_{0.05}$, the null hypothesis is rejected, as it would also be if the test level were $p = 0.01$ or 0.001.

8.   Therefore it is positively asserted that there is an association between the two sets of attributes, here sex and academic department, which cannot be attributed to sampling error.

9. The direction of the association is obtained by examining the pattern of observed and expected values, best seen when set out together in the contingency table. Here the direction of the association is that men are more likely than women to be studying science, women more likely than men to be studying arts, economics/social science and music.

It may be noted that the expected values have been calculated to an additional significant figure, compared with the observed values. This is appropriate for two reasons. First, the chi-square test is based on the assumption that the data consist of continuous variables, even though the observed values themselves are discrete. There is, therefore, no justification for restricting the expected values as if they were discrete variables. Second, a rounding of expected values to the same number of significant figures as the observed values does introduce an element of error into the χ^2 calculation which can be substantial, and reverse the result of the test when the calculated value is close to the critical value. Clearly, some discretion is needed in practice. This increased accuracy is likely to be more important, the smaller the general level of the expected values; it is not likely to be of practical importance if, for example, all the observed values have four digits. A fifth significant figure in the expected values would be most unlikely to affect the χ^2 calculation in any material way.

Note on 2 × 2 contingency tables

A chi-square test conducted on a 2×2 contingency table is exactly equivalent to a normal distribution test for the difference between two sample proportions where the samples are hypothesized to come from the same population, and a 'pooled estimate' from the two sample proportions is taken as the 'best estimate' of the population proportion (see section 5.9). So that

$$\mathrm{se}_{p_1 - p_2} = \sqrt{\left\{ p\,(1 - p)\left(\frac{1}{n_1} + \frac{1}{n_2}\right) \right\}}.$$

Consider two samples, size n_1 and n_2, in which the number of successes in each is x_1 and x_2. Then $p_1 = x_1/n_1$, $p_2 = x_2/n_2$ and $p = (x_1 + x_2)/(n_1 + n_2)$. If these data are then put in a 2×2 contingency table, the observed values, the sub-totals and grand total, from which can be derived the expected values, can all be expressed in terms of n_1, n_2, x_1 and x_2 as shown in Table 7.9.

Table 7.9 *A 2 × 2 contingency table*

	Sample 1	*Sample 2*	*Totals*
Success Failure	$o_1 = x_1$ $o_3 = n_1 - x_1$	$o_2 = x_2$ $o_4 = n_2 - x_2$	$x_1 + x_2$ $n_1 + n_2 - (x_1 + x_2)$
Totals	n_1	n_2	$n_1 + n_2$

It follows that

$$e_1 = \frac{(x_1 + x_2)\, n_1}{n_1 + n_2} = pn_1$$

Similarly $e_2 = pn_2$

$$e_3 = \frac{\{n_1 + n_2 - (x_1 + x_2)\}n_1}{n_1 + n_2} = (1 - p)n_1$$

Similarly, $e_4 = (1 - p)n_2$

$$o_1 - e_1 = x_1 - \frac{(x_1 + x_2)n_1}{n_1 + n_2} = \frac{n_2 x_1 - n_1 x_2}{n_1 + n_2}$$

$$= \left(\frac{x_1}{n_1} - \frac{x_2}{n_2}\right) \Big/ \left(\frac{1}{n_1} + \frac{1}{n_2}\right) = (p_1 - p_2) \Big/ \left(\frac{1}{n_1} + \frac{1}{n_2}\right)$$

Since the difference between each pair of observed and expected values is arithmetically the same in a 2 × 2 contingency table,

$$\sum_{i=1}^{4} \frac{(o_i - e_i)^2}{e_i} = (o_1 - e_1)^2 \left(\frac{1}{e_1} + \frac{1}{e_2} + \frac{1}{e_3} + \frac{1}{e_4}\right)$$

$$= \frac{(p_1 - p_2)^2}{\left(\frac{1}{n_1} + \frac{1}{n_2}\right)^2} \left\{\frac{1}{pn_1} + \frac{1}{pn_2} + \frac{1}{(1 - p)n_1} + \frac{1}{(1 - p)n_2}\right\}$$

$$= \frac{(p_1 - p_2)^2}{\left(\frac{1}{n_1} + \frac{1}{n_2}\right)^2} \left\{\frac{1}{p}\left(\frac{1}{n_1} + \frac{1}{n_2}\right) + \frac{1}{(1 - p)}\left(\frac{1}{n_1} + \frac{1}{n_2}\right)\right\}$$

$$= \frac{(p_1 - p_2)^2}{p(1 - p)\left(\frac{1}{n_1} + \frac{1}{n_2}\right)}.$$

This is the square of the normal distribution test formula for obtaining the z value. Also it can be seen from the chi-square tables that the values of chi-square for 1 degree of freedom are the square of the normal curve critical values for the same probabilities.

7.7   The contingency coefficient

Through the chi-square test, it is possible to establish whether the differences between the sets of observed and expected values in a contingency table are likely to be due to sampling error. If not, the differences can be looked upon as being statistically significant, so that some association exists between the distribution of the data over the two sets of attributes. The value of the χ^2 statistic does not indicate the strength of the association. An indication of this can be obtained from the contingency coefficient (C), which is given by

$$C = \sqrt{\left\{\frac{\chi^2}{\chi^2 + N}\right\}},$$

where N is the grand total of all frequencies in the table, so that it represents total sample size. As with the correlation coefficient, the larger the value of C, the stronger the association between the two sets of attributes. If there is no association, then $C = 0$; if the association is strong, C is close to 1. However, the maximum possible value of C varies according to the number of degrees of freedom, increasing as the number of degrees of freedom increases, subject to the limitation that the maximum value of C approaches close to, but can never actually reach 1. For example, for a 2×2 table with one degree of freedom, the maximum value of C is $0 \cdot 707$. For a 3×3 contingency table with four degrees of freedom, the maximum value of C is $0 \cdot 816$; and for a 4×4 contingency table with nine degrees of freedom, the maximum value of C is $0 \cdot 869$. The variation in maximum values of C with different degrees of freedom means that it is not strictly possible to compare the value of C between different contingency tables which do not have the same degree of freedom.

In the example in section 7.6 above, the calculated value of χ^2 was $20 \cdot 38$ and total sample size was 900. For these data, therefore,

$$C = \sqrt{\left(\frac{20 \cdot 38}{20 \cdot 38 + 900}\right)} = \sqrt{\left(\frac{20 \cdot 38}{920 \cdot 38}\right)} = 0 \cdot 140.$$

This is a low value for C, indicating that the association between sex and academic department, although statistically significant, is weak.

Exercises

(Section 7.4)

1.   A survey of absences due to sickness was conducted in a particular firm. During a ten-week period, staff were absent in aggregate for a total of 713 days. Absences were recorded by day of the week:

Monday	188	Thursday	124
Tuesday	141	Friday	108
Wednesday	152		

On the basis that the data are a random sample, is the apparent variation in sick absences by day of the week statistically significant? (Test the hypothesis that one fifth of all sick absences occur on each day of the week.)

2.   In a particular area, 138 persons were killed or seriously injured in road accidents over a twelve-month period. The monthly figures (adjusted for the length of each month) are as follows.

January	16·7	July	8·6
February	10·4	August	12·8
March	13·1	September	10·9
April	9·7	October	14·0
May	8·2	November	7·8
June	11·5	December	14·3

Assuming that this is a random sample, do the data establish that the number of persons killed or seriously injured in road accidents varies according to the season of the year? (Test the hypothesis that one-twelfth of all accidents occur in each month.)

(Section 7.6)

3.   From the data in Table 7.10 consider whether there is any difference between the parents of boys and girls in their attitudes to children staying at school after the minimum school-leaving age.

Table 7.10   *Parents wanting children to stay at secondary school after minimum school leaving age*

Parental attitude	Boys	Girls
Yes	3,345	3,132
No	176	169
Other and don't know	536	623
Total	4,057	3,924

Source: *First Report of the National Child Development Study*, 1966.

4. Using Table 7.11, test the hypothesis that parents' contact with head teacher is related to size of class.

Table 7.11   *Contact between parent and Head related to size of class*

Whether parent talked to Head when child started school or before	Size of class		
	Up to 30	31–40	Over 40
Talked to Head	41	79	42
Did not talk to Head	21	22	15
Total	62	101	57

Source: *Children and their Primary Schools*, Vol. II (Plowden Report), 1967.

5. In a particular survey of parental attitudes towards their children's schooling, equal-sized random samples of manual and non-manual parents of children in primary schools were asked whether they would prefer selection for secondary schools with a relatively small number of grammar places, or neighbourhood secondary schools to which all children would automatically transfer. The parents were asked to say whether (i) they preferred selection, (ii) were uncertain or thought there was little advantage in either system, or (iii) preferred non-selective neighbourhood schools. Out of a total of 84 non-manual parents, half (42) as compared with 26 of the manual parents favoured selection; more manual parents than non-manual

showed uncertainty about the respective advantages of selection and non-selection, but 45 manual parents compared with 34 non-manual clearly favoured non-selective neighbourhood schools.

Put the information contained in the above paragraph into a contingency table; then test whether there is any association between social class of parent and parental preferences in secondary schooling.

6. On the basis that the data in Table 7.12 are based on random samples, test whether there is any association between size of family in temporary accommodation and whether or not the family is fatherless. (Expected values to be calculated to whole numbers only.)

Table 7.12   *Number of families living in temporary accommodation at 31 December 1968*

	1 or 2 children	3 or more children	Total
Families with man, woman and child or children	849	1,712	2,561
Families with only woman and child or children	489	574	1,063
All families	1,338	2,286	3,624

Source: *Annual Report of the Department of Health and Social Security*, 1968.

7. Does Table 7.13 provide evidence to suggest that there is an association between the incidence of thumb-sucking and social class?

Table 7.13   *Thumb-sucking and social class*

	Social class					
	I & II	III (white collar)	III (manual)	IV	V	Totals
Thumb suckers	41	21	28	15	11	116
Non-thumb suckers	88	87	198	121	90	584
Totals	129	108	226	136	101	700

Source: J. & E. Newson—*Four Years Old in an Urban Community*, (Allen & Unwin, 1968).

K

8. Do the data in Table 7.14, which are derived from random samples, support the view that occupational group is related to promotion aspirations?

Table 7.14   *Promotion aspirations of three occupational groups*

Occupational group	Promotion aspirations		Total
	Liked the idea of promotion	*Did not like the idea of promotion*	
White-collar Workers	47	7	54
Skilled manual workers	49	30	79
Semi-skilled manual workers	65	85	150
Totals	161	122	283

Sources: J. H. Goldthorpe *et. al.*, 'The Affluent Worker and the Thesis of Embourgoisement', *Sociology*, Vol. 1, No. 1, January 1967.

9. On the assumption that the data in Table 7.15 are derived from random samples, consider whether there is an association between loneliness and marital status, with and without surviving children.

Table 7.15   *Loneliness among new residents at old peoples' homes: by marital status*

New residents saying they were	Marital status			All
	Unmarried	*Married, widowed and separated*		
		without surviving children	*with surviving children*	
Often lonely	18	21	52	91
Sometimes lonely	42	28	58	128
Not lonely	75	54	120	249
Totals	135	103	230	468

Source: Peter Townsend, *The Last Refuge* (Routledge & Kegan Paul, 1964).

Further reading

The chi-square distribution, contingency tables and tests for goodness of fit are discussed in:

A. Bradford Hill, *Principles of Medical Statistics* (The Lancet, 8th ed. 1966) chapters 13 and 14.
F. Conway, *Sampling, an Introduction for Social Scientists* (Allen and Unwin, 1967) chapter 4.
S. M. Dornbusch and C. F. Schmid, *A Primer of Social Statistics* (McGraw-Hill, 1955) chapters 14 and 15.
J. E. Freund and F. J. Williams, *Modern Business Statistics* (Pitman, 2nd ed. 1970) chapter 11.
M. J. Moroney, *Facts from Figures* (Penguin Books, 1951) chapter 15.
K. A. Yeomans, *Applied Statistics* (Penguin Books, 1968) chapter 6.

A broad approach, at a practical level, to tests involving contingency tables and goodness of fit is found in:

A. E. Maxwell, *Analysing Qualitative Data* (Methuen, 1961).

Appendix : Statistical tables
Table I Normal curve areas

z	0·00	0·01	0·02	0·03	0·04	0·05	0·06	0·07	0·08	0·09
0·0	0·0000	0·0040	0·0080	0·0120	0·0160	0·0199	0·0239	0·0279	0·0319	0·0359
0·1	0·0398	0·0438	0·0478	0·0517	0·0557	0·0596	0·0636	0·0675	0·0714	0·0753
0·2	0·0793	0·0832	0·0871	0·0910	0·0948	0·0987	0·1026	0·1064	0·1103	0·1141
0·3	0·1179	0·1217	0·1255	0·1293	0·1331	0·1368	0·1406	0·1443	0·1480	0·1517
0·4	0·1554	0·1591	0·1628	0·1664	0·1700	0·1736	0·1772	0·1808	0·1844	0·1879
0·5	0·1915	0·1950	0·1985	0·2019	0·2054	0·2088	0·2123	0·2157	0·2190	0·2224
0·6	0·2257	0·2291	0·2324	0·2357	0·2389	0·2422	0·2454	0·2486	0·2517	0·2549
0·7	0·2580	0·2611	0·2642	0·2673	0·2704	0·2734	0·2764	0·2794	0·2823	0·2852
0·8	0·2881	0·2910	0·2939	0·2967	0·2995	0·3023	0·3051	0·3078	0·3106	0·3133
0·9	0·3159	0·3186	0·3212	0·3238	0·3264	0·3289	0·3315	0·3340	0·3365	0·3389
1·0	0·3413	0·3438	0·3461	0·3485	0·3508	0·3531	0·3554	0·3577	0·3599	0·3621
1·1	0·3643	0·3665	0·3686	0·3708	0·3729	0·3749	0·3770	0·3790	0·3810	0·3830
1·2	0·3849	0·3869	0·3888	0·3907	0·3925	0·3944	0·3962	0·3980	0·3997	0·4015
1·3	0·4032	0·4049	0·4066	0·4082	0·4099	0·4115	0·4131	0·4147	0·4162	0·4177
1·4	0·4192	0·4207	0·4222	0·4236	0·4251	0·4265	0·4279	0·4292	0·4306	0·4319
1·5	0·4332	0·4345	0·4357	0·4370	0·4382	0·4394	0·4406	0·4418	0·4429	0·4441
1·6	0·4452	0·4463	0·4474	0·4484	0·4495	0·4505	0·4515	0·4525	0·4535	0·4545
1·7	0·4554	0·4564	0·4573	0·4582	0·4591	0·4599	0·4608	0·4616	0·4625	0·4633
1·8	0·4641	0·4649	0·4656	0·4664	0·4671	0·4678	0·4686	0·4693	0·4699	0·4706
1·9	0·4713	0·4719	0·4726	0·4732	0·4738	0·4744	0·4750	0·4756	0·4761	0·4767

z	0·00	0·01	0·02	0·03	0·04	0·05	0·06	0·07	0·08	0·09
2·0	0·4772	0·4778	0·4783	0·4788	0·4793	0·4798	0·4803	0·4808	0·4812	0·4817
2·1	0·4821	0·4826	0·4830	0·4834	0·4838	0·4842	0·4846	0·4850	0·4854	0·4857
2·2	0·4861	0·4864	0·4868	0·4871	0·4875	0·4878	0·4881	0·4884	0·4887	0·4890
2·3	0·4893	0·4896	0·4898	0·4901	0·4904	0·4906	0·4909	0·4911	0·4913	0·4916
2·4	0·4918	0·4920	0·4922	0·4925	0·4927	0·4929	0·4931	0·4932	0·4934	0·4936
2·5	0·4938	0·4940	0·4941	0·4943	0·4945	0·4946	0·4948	0·4949	0·4951	0·4952
2·6	0·4953	0·4955	0·4956	0·4957	0·4959	0·4960	0·4961	0·4962	0·4963	0·4964
2·7	0·4965	0·4966	0·4967	0·4968	0·4969	0·4970	0·4971	0·4972	0·4973	0·4974
2·8	0·4974	0·4975	0·4976	0·4977	0·4977	0·4978	0·4979	0·4979	0·4980	0·4981
2·9	0·4981	0·4982	0·4982	0·4983	0·4984	0·4984	0·4985	0·4985	0·4986	0·4986
3·0	0·49865	0·49869	0·49874	0·49878	0·49882	0·49886	0·49889	0·49893	0·49896	0·49900
3·1	0·49903	0·49906	0·49910	0·49913	0·49916	0·49918	0·49921	0·49924	0·49926	0·49929
3·2	0·4993129									
3·3	0·4995166									
3·4	0·4996631									
3·5	0·4997674									
3·6	0·4998409									
3·7	0·4998922									
3·8	0·4999277									
3·9	0·4999519									
4·0	0·4999683									
4·5	0·4999966									
5·0	0·4999997133									

Table II   Distribution of *t*

Degrees of freedom	Probability												
	0·9	0·8	0·7	0·6	0·5	0·4	0·3	0·2	0·1	0·05	0·02	0·01	0·001
1	0·158	0·325	0·510	0·727	1·000	1·376	1·963	3·078	6·314	12·706	31·821	63·657	636·619
2	0·142	0·289	0·445	0·617	0·816	1·061	1·386	1·886	2·920	4·303	6·965	9·925	31·598
3	0·137	0·277	0·424	0·584	0·765	0·978	1·250	1·638	2·353	3·182	4·541	5·841	12·924
4	0·134	0·271	0·414	0·569	0·741	0·941	1·190	1·533	2·132	2·776	3·747	4·604	8·610
5	0·132	0·267	0·408	0·559	0·727	0·920	1·156	1·476	2·015	2·571	3·365	4·032	6·869
6	0·131	0·265	0·404	0·553	0·718	0·906	1·134	1·440	1·943	2·447	3·143	3·707	5·959
7	0·130	0·263	0·402	0·549	0·711	0·896	1·119	1·415	1·895	2·365	2·998	3·499	5·408
8	0·130	0·262	0·399	0·546	0·706	0·889	1·108	1·397	1·860	2·306	2·896	3·355	5·041
9	0·129	0·261	0·398	0·543	0·703	0·883	1·100	1·383	1·833	2·262	2·821	3·250	4·781
10	0·129	0·260	0·397	0·542	0·700	0·879	1·093	1·372	1·812	2·228	2·764	3·169	4·587
11	0·129	0·260	0·396	0·540	0·697	0·876	1·088	1·363	1·796	2·201	2·718	3·106	4·437
12	0·128	0·259	0·395	0·539	0·695	0·873	1·083	1·356	1·782	2·179	2·681	3·055	4·318
13	0·128	0·259	0·394	0·538	0·694	0·870	1·079	1·350	1·771	2·160	2·650	3·012	4·221
14	0·128	0·258	0·393	0·537	0·692	0·868	1·076	1·345	1·761	2·145	2·624	2·977	4·140
15	0·128	0·258	0·393	0·536	0·691	0·866	1·074	1·341	1·753	2·131	2·602	2·947	4·076
16	0·128	0·258	0·392	0·535	0·690	0·865	1·071	1·337	1·746	2·120	2·583	2·921	4·015
17	0·128	0·257	0·392	0·534	0·689	0·863	1·069	1·333	1·740	2·110	2·567	2·898	3·965
18	0·127	0·257	0·392	0·534	0·688	0·862	1·067	1·330	1·734	2·101	2·552	2·878	3·922
19	0·127	0·257	0·391	0·533	0·688	0·861	1·066	1·328	1·729	2·093	2·539	2·861	3·883
20	0·127	0·257	0·391	0·533	0·687	0·860	1·064	1·325	1·725	2·086	2·528	2·845	3·850

Degrees of freedom						Probability							
	0·9	0·8	0·7	0·6	0·5	0·4	0·3	0·2	0·1	0·05	0·02	0·01	0·001
21	0·127	0·257	0·391	0·532	0·686	0·859	1·063	1·323	1·721	2·080	2·518	2·831	3·819
22	0·127	0·256	0·390	0·532	0·686	0·858	1·061	1·321	1·717	2·074	2·508	2·819	3·792
23	0·127	0·256	0·390	0·532	0·685	0·858	1·060	1·319	1·714	2·069	2·500	2·807	3·767
24	0·127	0·256	0·390	0·531	0·685	0·857	1·059	1·318	1·711	2·064	2·492	2·797	3·745
25	0·127	0·256	0·390	0·531	0·684	0·856	1·058	1·316	1·708	2·060	2·485	2·787	3·725
26	0·127	0·256	0·390	0·531	0·684	0·856	1·058	1·315	1·706	2·056	2·479	2·779	3·707
27	0·127	0·256	0·389	0·531	0·684	0·855	1·057	1·314	1·703	2·052	2·473	2·771	3·690
28	0·127	0·256	0·389	0·530	0·683	0·855	1·056	1·313	1·701	2·048	2·467	2·763	3·674
29	0·127	0·256	0·389	0·530	0·683	0·854	1·055	1·311	1·699	2·045	2·462	2·756	3·659
30	0·127	0·256	0·389	0·530	0·683	0·854	1·055	1·310	1·697	2·042	2·457	2·750	3·646
40	0·126	0·255	0·388	0·529	0·681	0·851	1·050	1·303	1·684	2·021	2·423	2·704	3·551
60	0·126	0·254	0·387	0·527	0·679	0·848	1·046	1·296	1·671	2·000	2·390	2·660	3·460
120	0·126	0·254	0·386	0·526	0·677	0·845	1·041	1·289	1·658	1·980	2·358	2·617	3·373
∞	0·126	0·253	0·385	0·524	0·674	0·842	1·036	1·282	1·645	1·960	2·326	2·576	3·291

This table is taken from Table III of Fisher and Yates: *Statistical Tables for Biological, Agricultural and Medical Research* (6th ed. 1963), published by Oliver and Boyd, Edinburgh, and by permission of the authors and publishers.

Probability here refers to the proportion of total area in the two tails of the t distribution, and thus provides a suitable subscript for two-tailed tests. For example, at six degrees of freedom, $t_{0.05} = 2.447$. For one-tailed tests, the probability is *halved* for any given t value. For a one-tailed test at six degrees of freedom, $t_{0.05} = 1.943$ or $t_{0.025} = 2.447$.

Table III   Distribution of chi-square

Degrees of freedom	Probability											
	0·95	0·90	0·80	0·70	0·50	0·30	0·20	0·10	0·05	0·02	0·01	0·001
1	0·00393	0·0158	0·0642	0·148	0·455	1·074	1·642	2·706	3·841	5·412	6·635	10·827
2	0·103	0·211	0·446	0·713	1·386	2·408	3·219	4·605	5·991	7·824	9·210	13·815
3	0·352	0·584	1·005	1·424	2·366	3·665	4·642	6·251	7·815	9·837	11·345	16·266
4	0·711	1·064	1·649	2·195	3·357	4·878	5·989	7·779	9·488	11·668	13·277	18·467
5	1·145	1·610	2·343	3·000	4·351	6·064	7·289	9·236	11·070	13·388	15·086	20·515
6	1·635	2·204	3·070	3·828	5·348	7·231	8·558	10·645	12·592	15·033	16·812	22·457
7	2·167	2·833	3·822	4·671	6·346	8·383	9·803	12·017	14·067	16·622	18·475	24·322
8	2·733	3·490	4·594	5·527	7·344	9·524	11·030	13·362	15·507	18·168	20·090	26·125
9	3·325	4·168	5·380	6·393	8·343	10·656	12·242	14·684	16·919	19·679	21·666	27·877
10	3·940	4·865	6·179	7·267	9·342	11·781	13·442	15·987	18·307	21·161	23·209	29·588
11	4·575	5·578	6·989	8·148	10·341	12·899	14·631	17·275	19·675	22·618	24·725	31·264
12	5·226	6·304	7·807	9·034	11·340	14·011	15·812	18·549	21·026	24·054	26·217	32·909
13	5·892	7·042	8·634	9·926	12·340	15·119	16·985	19·812	22·362	25·472	27·688	34·528
14	6·571	7·790	9·467	10·821	13·339	16·222	18·151	21·064	23·685	26·873	29·141	36·123
15	7·261	8·547	10·307	11·721	14·339	17·322	19·311	22·307	24·996	28·259	30·578	37·697
16	7·962	9·312	11·152	12·624	15·338	18·418	20·465	23·542	26·296	29·633	32·000	39·252
17	8·672	10·085	12·002	13·531	16·338	19·511	21·615	24·769	27·587	30·995	33·409	40·790
18	9·390	10·865	12·857	14·440	17·338	20·601	22·760	25·989	28·869	32·346	34·805	42·312
19	10·117	11·651	13·716	15·352	18·338	21·689	23·900	27·204	30·144	33·687	36·191	43·820
20	10·851	12·443	14·578	16·266	19·337	22·775	25·038	28·412	31·410	35·020	37·566	45·315

Degrees of freedom					Probability							
	0·95	0·90	0·80	0·70	0·50	0·30	0·20	0·10	0·05	0·02	0·01	0·001
21	11·591	13·240	15·445	17·182	20·337	23·858	26·171	29·615	32·671	36·343	38·932	46·797
22	12·338	14·041	16·314	18·101	21·337	24·939	27·301	30·813	33·924	37·659	40·289	48·268
23	13·091	14·848	17·187	19·021	22·337	26·018	28·429	32·007	35·172	38·968	41·638	49·728
24	13·848	15·659	18·062	19·943	23·337	27·096	29·553	33·196	36·415	40·270	42·980	51·179
25	14·611	16·473	18·940	20·867	24·337	28·172	30·675	34·382	37·652	41·566	44·314	52·620
26	15·379	17·292	19·820	21·792	25·336	29·246	31·795	35·563	38·885	42·856	45·642	54·052
27	16·151	18·114	20·703	22·719	26·336	30·319	32·912	36·741	40·113	44·140	46·963	55·476
28	16·928	18·939	21·588	23·647	27·336	31·391	34·027	37·916	41·337	45·419	48·278	56·893
29	17·708	19·768	22·475	24·577	28·336	32·461	35·139	39·087	42·557	46·693	49·588	58·302
30	18·493	20·599	23·364	25·508	29·336	33·530	36·250	40·256	43·773	47·962	50·892	59·703

This table is taken from Table IV of Fisher and Yates: *Statistical Tables for Biological, Agricultural and Medical Research* (6th ed. 1963), published by Oliver and Boyd, Edinburgh, and by permission of the authors and publishers.

Answers to Exercises

Chapter 1

1. (a) 1/2, (b) 1/4, (c) 1/13, (d) 1/26.
2. (a) 5/13, (b) 2/13, (c) 4/13.
3. (a) 1/4, (b) 1/16, (c) 9/169, (d) 1/2.
4. (a) 25/102, (b) 1/17, (c) 11/221, (d) 26/51.
5. $0 \cdot 1$.
6. (a) $0 \cdot 12$, (b) $0 \cdot 58$.
7. $0 \cdot 62$.
9. Probabilities are 4/7 for A, 3/7 for B.

Chapter 2

1. (a) $2 \cdot 5$, (b) $-2 \cdot 8$, (c) $11 \cdot 1$, (d) $-0 \cdot 2$, (e) $4 \cdot 3$, (f) $-7 \cdot 5$.
2. (a) $0 \cdot 1587$, (b) $0 \cdot 9608$, (c) $0 \cdot 8023$, (d) $0 \cdot 9861$, (e) $0 \cdot 5102$, (f) $0 \cdot 1378$, (g) $0 \cdot 2951$.
3. (a) $1 \cdot 16$, (b) $-1 \cdot 22$, (c) $0 \cdot 2$, (d) $0 \cdot 43$, (e) $-2 \cdot 54$, (f) $1 \cdot 29$.
4. (a) 251, (b) 877, (c) $36 \cdot 25$ years, (d) $61 \cdot 32$ years.
5. (a) $0 \cdot 17$, (b) $0 \cdot 02$, (c) $0 \cdot 002$, (d) $0 \cdot 62$.
6. (a) $0 \cdot 13$, (b) $0 \cdot 28$, (c) $0 \cdot 55$, (d) $0 \cdot 01$.

Chapter 4

1. $se_{\bar{x}} = £0 \cdot 28$; 95% c.l., £21·70 to £22·80; 99% c.l., £21·53 to £22·98.
2. $se_{\bar{x}} = 0 \cdot 28$ years; 95% c.l., 50·52 to 51·62 years; 99% c.l., 50·35 to 51·79 years.
3. (a) $\bar{x} = 1 \cdot 405$ workers, $s = 1 \cdot 024$ workers; (b) $se_{\bar{x}} = 0 \cdot 012$ workers; 95% c.l., 1·381 to 1·429 workers.
4. $se_{\bar{x}} = 0 \cdot 152$ cm, $0 \cdot 30$ cm.
5. $se_{\bar{x}} = 0 \cdot 018$ persons; 0·95 probable error is 0·035 persons, 0·99 probable error is 0·046 persons.
6. $se_{\bar{x}} = £2 \cdot 74$; 95% c.l., £34·44 to £45·76; 99% c.l., £32·42 to £47·78.
7. $se_{\bar{x}} = £0 \cdot 504$; £1·04.

8. $\text{se}_p = 0\cdot0325\ (3\cdot25\%)$; 99% c.l., $0\cdot796$ to $0\cdot964$ (79·6% to 96·4%).
9. $\text{se}_p = 0\cdot0302\ (3\cdot02\%)$; $0\cdot95$ probable error is $0\cdot0592$ $(5\cdot92\%)$.
10. $\text{se}_p = 0\cdot0068\ (0\cdot68\%)$; $0\cdot95$ probable error is $0\cdot0134$ $(1\cdot34\%)$; $0\cdot99$ probable error is $0\cdot0176\ (1\cdot76\%)$.
11. $\text{se}_p = 0\cdot0203\ (2\cdot03\%)$; 95% c.l., $0\cdot201$ to $0\cdot290$ (21·0% to 29·0%); 99% c.l., $0\cdot198$ to $0\cdot302$ (19·8% to 30·2%).
12. $\text{se}_p = 0\cdot0056\ (0\cdot56\%)$; 99·9% c.l., $0\cdot637$ to $0\cdot674$ (63·7% to 67·4%).

Chapter 5

1. $\text{se}_{\bar{x}} = 0\cdot24$ cm; $z = 8\cdot33$.
2. $\text{se}_{\bar{x}} = 0\cdot89$; $z = -1\cdot68$.
3. $\bar{x} = 47\cdot24$ years; $s = 10\cdot86$ years; $\text{se}_{\bar{x}} = 0\cdot26$ years; $z = 2\cdot9$.
4. $\text{se}_{\bar{x}} = £0\cdot75$; $t = -3\cdot47$.
5. $\text{se}_{\bar{x}_1 - \bar{x}_2} = £0\cdot14$; $z = 3\cdot96$.
6. $\bar{x}_1 = 3\cdot06$ persons; $s_1 = 1\cdot50$ persons; $\bar{x}_2 = 2\cdot86$ persons, $s_2 = 1\cdot37$ persons; $\text{se}_{\bar{x}_1 - \bar{x}_2} = 0\cdot096$ persons; $z = 2\cdot08$.
7. $\text{se}_{\bar{x}_1 - \bar{x}_2} = 0\cdot56$ cm; $z = 1\cdot79$.
8. $\text{se}_{\bar{x}_1 - \bar{x}_2} = £0\cdot24$; $t = 1\cdot84$.
9. $\text{se}_p = 0\cdot0218\ (2\cdot18\%)$; $z = 0\cdot92$.
10. $\text{se}_p = 0\cdot0120\ (1\cdot20\%)$; $z = 1\cdot59$.
11. $\text{se}_p = 0\cdot0344\ (3\cdot44\%)$; $z_1 = -0\cdot87$, $z_2 = 3\cdot49$.
12. $p_1 = 0\cdot61\ (61\%)$; $p_2 = 0\cdot59\ (59\%)$; $p = 0\cdot602\ (60\cdot2\%)$; $\text{se}_{p_1 - p_2} = 0\cdot0316\ (3\cdot16\%)$; $z = 0\cdot63$.
13. (a) $p = 0\cdot554\quad(55\cdot4\%)$; $\text{se}_{p_1 - p_2} = 0\cdot0700\quad(7\cdot00\%)$; $z = 1\cdot71$.
    (b) $p = 0\cdot198\quad(19\cdot8\%)$; $\text{se}_{p_1 - p_2} = 0\cdot0500\quad(5\cdot00\%)$; $z = -2\cdot00$.
14. (a) $p = 0\cdot155\quad(15\cdot5\%)$; $\text{se}_{p_1 - p_2} = 0\cdot0076\quad(0\cdot76\%)$; $z = -2\cdot89$.
    (b) $p = 0\cdot0615\quad(6\cdot15\%)$; $\text{se}_{p_1 - p_2} = 0\cdot0026\quad(0\cdot26\%)$; $z = 3\cdot08$.
15. $\text{se}_{\bar{x}} = £0\cdot14$; $z = -6\cdot29$.
16. $\text{se}_{\bar{x}} = £1\cdot01$; $z = 2\cdot68$.
17. $p_1 = 0\cdot590\ (59\cdot0\%)$; $p_2 = 0\cdot714\ (71\cdot4\%)$; $p = 0\cdot659$ $(65\cdot9\%)$; $\text{se}_{p_1 - p_2} = 0\cdot0318\ (3\cdot18\%)$; $z = -3\cdot90$.

Chapter 6

1. $\text{se}_{\bar{x}} = £0\cdot302$; $z = -3\cdot07$.

2. $se_{\bar{x}} = 0 \cdot 209$ years; $0 \cdot 975$ probable minimum increase is $0 \cdot 39$ years.
3. $se_{\bar{x}} = 0 \cdot 132$ months; $0 \cdot 995$ probable minimum increase is $1 \cdot 06$ months.
4. $se_{\bar{x}_1 - \bar{x}_2} = £0 \cdot 207$; $0 \cdot 975$ probable minimum difference is $£0 \cdot 32$; $0 \cdot 975$ probable maximum difference is $£1 \cdot 14$.
5. $se_{\bar{x}_1 - \bar{x}_2} = 0 \cdot 216$ cm; $z = -6 \cdot 48$; $0 \cdot 995$ probable minimum difference is $0 \cdot 84$ cm.
6. $p = 0 \cdot 112$ $(11 \cdot 2\%)$; $se_p = 0 \cdot 0083$ $(0 \cdot 83\%)$; $0 \cdot 975$ probable maximum difference is $0 \cdot 0393$ $(3 \cdot 93\%)$; $0 \cdot 975$ probable minimum difference is $0 \cdot 0067$ $(0 \cdot 67\%)$.
7. $se_p = 0 \cdot 0126$ $(1 \cdot 26\%)$; $0 \cdot 995$ probable minimum increase is $0 \cdot 0275$ $(2 \cdot 75\%)$.
8. $se_{p_1 - p_2} = 0 \cdot 0207$ $(2 \cdot 07\%)$; $z = -2 \cdot 42$; $0 \cdot 975$ probable minimum difference is $0 \cdot 0094$ $(0 \cdot 94\%)$.
9. $se_{p_1 - p_2} = 0 \cdot 0275$ $(2 \cdot 75\%)$; $z = 2 \cdot 55$.
10. $p_1 = 0 \cdot 784$ $(78 \cdot 4\%)$; $p_2 = 0 \cdot 617$ $(61 \cdot 7\%)$; $se_{p_1 - p_2} = 0 \cdot 0238$ $(2 \cdot 38\%)$; $0 \cdot 975$ probable difference is $0 \cdot 1203$ $(12 \cdot 03\%)$.

Chapter 7

1. $\chi^2 = 25 \cdot 9$ (4 degrees of freedom)
2. $\chi^2 = 7 \cdot 23$ (11 d.f.)
3. $\chi^2 = 10 \cdot 97$ (2 d.f.)
4. $\chi^2 = 2 \cdot 83$ (2 d.f.)
5. $\chi^2 = 6 \cdot 50$ (2 d.f.)
6. $\chi^2 = 54$ (1 d.f.)
7. $\chi^2 = 30 \cdot 2$ (4 d.f.)
8. $\chi^2 = 32 \cdot 2$ (2 d.f.)
9. $\chi^2 = 8 \cdot 39$ (4 d.f.)

Index